Fish locomotion

Fish locomotion

R. W. BLAKE

Department of Zoology, University of British Columbia, Vancouver, Canada

CAMBRIDGE UNIVERSITY PRESS

Cambridge

London   New York   New Rochelle

Melbourne   Sydney

Published by the Press Syndicate of the University of Cambridge
The Pitt Building, Trumpington Street, Cambridge CB2 1RP
32 East 57th Street, New York, NY 10022, USA
296 Beaconsfield Parade, Middle Park, Melbourne 3206, Australia

© Cambridge University Press 1983

First published 1983

Printed in Great Britain by the University Press, Cambridge

Library of congress catalogue card number: 82-12861

British Library cataloguing in publication data

Blake, R. W.
Fish locomotion.
1. Fishes—Locomotion
I. Title
597'.01'852   QL639.4

ISBN 0 521 24303 3

This book is dedicated to the memory of
my grandfather, William Deighton

Contents

Preface

This book is about the kinematics, mechanics and energetics of fish swimming. Interest in fish locomotion dates back to ancient times. In Europe the first recorded references to it are due to Aristotle (e.g. *The movement of animals*, fourth century B.C.). Aristotle implied that the pectoral and pelvic fins are used for swimming in most fish and that forms without these fins are propelled through the water in the same way that a snake moves over the ground. We now know that most fish are propelled by undulatory body waves that pass backwards along the body. However, it would seem that Aristotle anticipated Newton by 20 centuries in realizing the physical basis of the process. There seems to have been an early interest in fish swimming in Asia too. It is thought that references in the ancient Hindu work *Sustruta-samhita* to the correlation between body form, habitat and locomotion in certain freshwater fish predate the works of Aristotle (Hora, 1935).

Aristotle's notion that most fish row with their paired fins was dispelled by Giovanni Borelli in his book *Du motu animalium* which was published in 1680. Borelli was a student of Galileo and was well versed in the mechanics of his day. He seems to have realized that the paired fins are employed in braking and for initiating manoeuvres. However, his explanation of the mechanics of the caudal fin was based on an analogy with the 'side-to-side' movement of an oar and we now know this to be incorrect. Considering that there was no theory of undulatory waves or experimental techniques available for their study at this time, the limitations of Borelli's approach can be understood. Almost a century later, Pettigrew believed that the body of the fish always forms an integral number of standing sine waves. This is not the case.

The late nineteenth century saw the introduction of cinematography. E. J. Marey was a pioneer of this technique and of its application to the study of animal locomotion. Marey (1894) filmed a variety of fish including the common eel *Anguilla*. He was able to show that the undulatory waves generated by the fish pass back down its body at a speed greater than the forward swimming speed.

By the early twentieth century, engineers had begun to develop a quanti-

ix

tative approach to fish swimming problems. For example, Houssay (1912) and Magnan (1930) attempted to calculate the power output of swimming fish by measuring the velocity at which they swam against different loads. Shoulejkin (1929) employed a wind tunnel to measure the lift and drag forces acting on the pectoral fins and body of the flying fish *Exocoetus*. Kempf & Neu (1932) measured the drag force acting on a dead pike (*Esox*) in a large nautical towing tank. At about the same period Breder (1926) published a review of the fish locomotion literature of the time and coined many of the descriptive terms that are still used today to distinguish different swimming styles (e.g. anguilliform, carangiform, ostraciiform).

The basis of much of our current understanding of the kinematics and mechanics of animal locomotion is due to the work of Sir James Gray and his colleagues. Most of Gray's early work on animal locomotion was concerned with fish swimming. In addition to using cinematography to study undulatory propulsion, Gray designed a machine which could generate a given wave motion onto the body of a dead fish. The fish was impaled by a series of rods which were connected together by cams which could be adjusted to vary the phase difference between the rods, enabling a variety of waveforms to be generated. Information derived from these techniques was employed in a hydromechanical model of swimming based on the assumption that the process could be understood by describing the transverse motions of a series of arbitrarily defined segments.

Gray is perhaps best remembered for his work on the propulsive performance of the dolphin (Gray, 1936). Using the standard hydrodynamic equations of resistance and observations of swimming speed, the power required for swimming could be calculated. Gray equated this with a calculation of the power available from the propulsive musculature. The calculation was based on the assumption that a unit mass of dolphin muscle is capable of the same power output as that of a man performing strenuous activity. A discrepancy between the power required and that available of a factor of about seven was found. This result stimulated interest in the possibility of morphological drag reduction mechanisms in dolphins and fish and the likelihood that not all vertebrate striated muscles produce the same power per unit weight.

At about the same time that Gray was investigating the kinematics of undulatory propulsion, a colleague of his, J. E. Harris, was working on problems of stability and control in fish (Harris, 1935, 1936). Like Shoulejkin, Harris employed a wind tunnel and model fish in his experiments. In addition to this, Harris was one of the first zoologists to consider the diversity of form in fish in relation to their locomotor function. Harris (1937a,b, 1953) gave the first detailed descriptions of the kinematics of undulatory and oscillatory median and paired propulsion in fish.

Stimulated by the work of Gray and his colleagues, Sir Geoffrey Taylor developed the first hydromechanical model of swimming appropriate to the undulatory motions of fish (Taylor, 1952). Like Gray, Taylor assumed that the animal could be likened to a series of short segments that functioned independently of each other. The model was designed to calculate the normal and tangential resistive forces acting on the segments. These forces are dependent on the instantaneous velocity of the fish relative to the water. Resistive theories of undulatory swimming were developed initially for application to organisms smaller than typical adult fish (i.e. for low Reynolds Numbers) and therefore do not take account of any reactive (inertial) forces.

More recently, reactive theories of fish swimming have been developed that emphasize inertial effects and are therefore more appropriate for the high Reynolds Number situations typical of most fish. In contrast to the earlier resistive models, the new reactive theories emphasize forces that are dependent on the rate of change of the relative velocity of the segments. Much of the work in this area is due to Sir James Lighthill and T. Y. Wu (e.g. Lighthill, 1960, 1970, 1971; Wu, 1961, 1971a,b,c,d). Lighthill's work has been particularly influential. In his book *Mathematical biofluiddynamics* (1975), Lighthill summarizes his work and clearly shows the value of employing reactive theories (e.g. elongated-body theory) in explaining the hydro-mechanical significance of morphological adaptation in fish.

Lighthill (1971) applied elongated-body theory to a series of kinematic data on swimming dace (*Leuciscus*) that had been gathered by R. Bainbridge some years earlier (Bainbridge, 1963). Lighthill was able to show that the thrust required to swim was about four times greater than would be required to propel the equivalent rigid body at the same speed. A possible explanation for this was suggested by Q. Bone. Bone (in Lighthill, 1971) argues that an increase in frictional drag occurs in an oscillating body over its rigid equivalent due to the influence of boundary layer thinning. This will be explained further in chapter 5. Since the early 1970's, Lighthill's theory has been central to experimental work aimed at determining thrust and power requirements in swimming fish (e.g. Webb, 1971a,b; Wardle & Reid, 1977).

Much of the work described briefly above that was carried out during and subsequent to Gray's time is discussed at length in some of the chapters that follow. A more detailed account of the historical aspects may be found in Alexander (1983) however.

This work aims to build on previous syntheses, such as P. W. Webb's (1975a) review 'Hydrodynamics and energetics of fish propulsion' and recent symposia (e.g. *Swimming and flying in nature* (Wu, Brokaw & Brennan, 1974) and *Scale effects in animal locomotion* (Pedley, 1977)). A chapter-by-chapter outline of the main issues and topics covered in this book is given below. Much of the material covered is relatively recent (post-1975) and is not reviewed in any previous synthesis.

Chapter 1 reviews some of the principal physical concepts that are required to understand the chapters that follow. It begins with basic definitions of fluids and their physical properties, and moves on to consider the origin of drag and lift on the basis of the boundary layer concept. Readers may find an introductory fluid mechanics text, such as A. Shapiro's *Shape and flow* (Shapiro, 1964), a useful supplement to the material covered here.

The mechanics of swimming muscle is discussed in chapter 2. After reviewing the basic force–velocity properties of vertebrate striated muscle, emphasis is given to muscle fibre types and their role in swimming. Muscle-fibre anatomy and geometry is also discussed. This chapter is not intended to be a review of the structure and properties of muscle. Rather, it is designed to give a reasonable background to the mechanical function of the fish's power plant. References are given to general reviews that cover other aspects of muscle function (e.g. biochemical and neural aspects).

In chapter 3 swimming speed is related to size and power requirements. After considering definitions of levels of swimming activity and speed and stamina, swimming performance is discussed in the context of Gray's Paradox. The historical development of the ideas concerning power requirements relative to power availability is outlined and related to observations of swimming speed. Taking account of recent developments in muscle physiology and hydromechanical theory, new expressions relating predicted swimming speed to length are derived for various activity levels and boundary layer flow conditions. It is concluded that Gray's Paradox only applies to a few limited cases.

Drag and drag reduction mechanisms are discussed in chapter 4. Firstly, the methods of measuring drag forces are reviewed; then the validity of the so-called 'rigid-body analogy' (that swimming drag and the drag of the equivalent rigid body are the same) is discussed for streamlined and un-streamlined fish. Drag reduction mechanisms in fish are described and their likely effectiveness is evaluated in relation to mode of life. It is interesting to note that research into drag reduction mechanisms was prompted by Gray's Paradox. It can be argued that the value of the so-called paradox lies not in its contribution to our understanding of fish swimming mechanics as such, but in its indirect influence in stimulating research into drag reduction mechanisms and other issues.

Chapter 5 is concerned with the kinematics and mechanics of undulatory swimming. A basic distinction is made between steady (time-independent) and unsteady (time-dependent) swimming. Steady swimming modes (e.g. anguilliform, carangiform) are described. Swimming kinematics (e.g. waveform wavelength, amplitude and frequency) are discussed in relation to speed and size. The kinematics of fast-starts and turns are considered under the heading of unsteady swimming. The second half of chapter 5 is concerned with hydromechanical models of swimming. Both resistive and reactive models of undulatory propulsion are outlined. Emphasis is given to reactive theory, in particular to Lighthill's elongated-body theory, which forms the basis of models of steady and unsteady swimming mechanics. The theory is employed in predicting optimal morphologies and kinematics for undulatory swimming.

Median and paired fin propulsion is not as well understood at this time as undulatory swimming involving the body and caudal fin. In chapter 6 our knowledge of the so-called 'non-body' modes of swimming is reviewed. Again, emphasis is placed on employing hydromechanical theory to gain some understanding of the morphological design of fish. Two broad categories of non-body swimmers are identified: those propelled by undulating fins and those that employ oscillatory mechanisms. Undulatory fins are modelled on the basis of momentum-jet and slender-body theory. Oscillatory mechanisms can be further subdivided into drag- and lift-based types. Both are analysed employing blade-element theory. Results from the analyses are discussed in relation to mode of life and compared to findings on forms that swim in the undulatory body modes.

Like the mechanics of paired and median fin propulsion, stability and control has been a neglected area in fish locomotion research. However, there have been a few detailed studies on sharks and tuna and these are discussed in chapter 7. The function of the heterocercal tail in the sharks is also discussed, with emphasis on its role in swimming equilibrium.

One of the most exciting areas of research in fish swimming to develop over the past five years or so concerns swimming strategy and this is discussed in chapter 8. Optimal cruising for maximum energetic advantage, the possible advantages of intermittent (burst-and-glide) swimming over continuous steady swimming and energetically efficient swimming strategies for negatively buoyant fish are among the topics considered. In chapter 9 a variety of topics that do not fall naturally into any of the previous chapter headings are discussed; for example, swimming in larval fish, which requires a different approach to that emphasized in chapter 5 due to the low Reynolds Numbers involved.

This book is intended to be a summary for researchers in biomechanics and assumes a working knowledge of mathematics and biology. However, it is hoped that undergraduates may find it a useful reference text for upper-level courses in biomechanics. I would like to point out that certain parts of this book owe a great deal to the original research efforts of particular individuals. For example, much of the theoretical and experimental material covered in chapter 5 is due to Professor Sir James Lighthill F.R.S. and Dr P. W. Webb, respectively, and most of the work discussed in chapter 8 is the result of research by Professor D. Weihs. In addition to these people I would like to express my thanks to Professor R. McN. Alexander, Dr R. Bainbridge, Dr Q. Bone, Dr C. P. Ellington, Dr J. M. Gosline, Dr K. E. Machin, Professor N. B. Marshall, Dr C. C. Lindsey, Dr H. W. Lissman F.R.S., Dr C. J.

Pennycuick, Dr J. J. Videler and Dr C. Wardle for discussions on fish swimming. I would like to thank my wife, Pat, for her support and help on this project. I am grateful to the Natural and Engineering Research Council of Canada and the University of British Columbia for financial support. Finally, it is a pleasure to thank Mr M. Walters and all at the Cambridge University Press that were involved in the production of this book.

R. W. B.
Vancouver, 1982

Notation

The more important symbols are defined below. In some cases convention has necessitated the use of one symbol for more than one parameter and secondary meanings appear in brackets. All symbols are defined in context upon their first appearance in the text and it is thought that no serious ambiguities will arise. Barred symbols in the text refer to average (mean) quantities.

s	distance (span)
x	distance from the leading edge of a plate or streamlined body
d	diameter (distance between two plates)
l	length
b	breadth of a plate
c	chord
r	distance out from the base of a fin
R	fin length
J	height of a fish above the substrate
Y	stride length
δ	boundary layer thickness
t	time (thickness)
t_p	time of pectoral fin power stroke
t_r	time of pectoral fin recovery stroke
t_0	total cycle time
A	frontally projected area (amplitude)
S_w	total wetted surface area
S_d	actuator-disc area
m	mass
M	virtual mass
M_w	muscle weight
$(mg)_w$	submerged weight of a fish
U	forward velocity
U_{crit}	critical swimming speed
U_{max}	maximum swimming speed

U_o	optimum steady swimming speed
U_{min}	minimum swimming speed
U_e	maximum sustainable swimming speed
U_f	final speed at the termination of a glide
U_{slip}	slipping speed of a negatively buoyant fish
U_{cross}	'crossover speed'
U_a	speed of a flying fish relative to the air
U_w	wind speed
u'	tidal speed
v	instantaneous velocity
v_n	normal velocity component in pectoral fin power stroke
v_n'	normal component of fin velocity in pectoral fin recovery stroke
v_s	spanwise velocity component in pectoral fin power stroke
v_s'	spanwise velocity component in pectoral fin recovery stroke
v_{res}	resultant velocity component in pectoral fin power stroke
v_f	flapping velocity of a pectoral fin
W	lateral velocity of a body segment (work)
W_r	relative velocity of a body segment
w	lateral velocity of 'pushing' of a body segment (tangential velocity component in slender-body theory)
V	body-wave velocity (velocity of muscular shortening)
ω	angular velocity
a	acceleration (constant in force–velocity equation of muscular shortening)
v_y'	dv_y/dt, where v_y is a velocity component in the y direction
v_x'	dv_x/dt, where v_x is a velocity component in the x direction
λ_s	oscillatory wavelength of a body segment
λ_b	wavelength of body wave
f	waveform frequency
f_{min}	length-dependent minimum waveform frequency
γ	positional angle of a fin
θ_d	downwash angle
θ_s	angle subtended by a body segment relative to its axis of progression (c, d)
θ_b	angle of body relative to the axis of progression (c, d)
θ_1	'glide down' angle of a negatively buoyant fish
θ_2	'ascent angle' of a negatively buoyant fish
α	angle of attack
Λ	geometrical angle of attack
θ_e	emergence angle
μ	viscosity
ν	kinematic viscosity
ρ	density

xvi

F	force
D_f	frictional drag force
D	drag force
L	lift force
L_{tot}	total lifting force
$F_{(v)}$	vertically directed force
$F_{(h)}$	horizontally directed force
F_n	normal force component acting on a pectoral fin blade-element during its power stroke
F_n'	normal force component acting on a pectoral fin blade-element during its recovery stroke
F_s	spanwise force acting on a pectoral fin blade-element
F_c	chordwise force acting on a pectoral fin blade-element
F_a	added mass force
F_b	force acting on the body of a fish
T	thrust force
(T, F_s)	reactive force in slender-body theory
p	pressure
p_0	static pressure
τ	shear stress
I	impulse (intensity of turbulence)
W_T	total work
W_K	kinetic energy
W_P	potential energy
W_p	work required to overcome profile drag
W_i	inertial work
W_a	work associated with an added mass force
W_b	work required to drag the body of a fish
W_c	work expended in steady swimming at a constant velocity
W_g	work expended in 'burst-and-glide' swimming
W_{tp}	work expended in tidal transport swimming
W_{int}	energy expended in intermittent swimming of a negatively buoyant fish
W_j	work expended in jumping
$W_{resistive}$	work associated with the resistive force generated by a segment of the body of a larval fish
W_{head}	work required to move the head of a larval fish
W_c^*	energy expenditure per unit distance for a fish swimming at constant speed
W_g^*	energy expenditure per unit distance for burst-and-glide swimming
S	a given energy store
Q	torque

Q_p	torque associated with the profile drag acting on a pectoral fin
Q_i	inertial torque
Q_a	torque associated with the added mass of a pectoral fin
P	power (force in the force–velocity equation for a shortening muscle)
P_T	total power
P_K	power associated with kinetic energy
P_{req}	power required
P_{avail}	power available
P_{in}	power input
P_{id}	induced power
P_b	power required to overcome the drag acting on the body of a fish
P_p	power required to overcome the profile drag of a pectoral fin
P_i	power required to overcome fin inertia
P_a	power associated with an added mass force
P_m	metabolic power
P_s	swimming power
P_f	power factor
g	gravitational constant
C_f	frictional drag coefficient
$C_{f(lam)}$	frictional drag coefficient based on a laminar boundary layer
$C_{f(turb)}$	frictional drag coefficient based on a turbulent boundary layer
$C_{f(tran)}$	frictional drag coefficient based on a transitional boundary layer
C_p	pressure drag coefficient
C_D	drag coefficient
$C_{D(b)}$	drag coefficient of the body of a fish
C_{Do}	profile drag coefficient
C_{Di}	induced drag coefficient
C_L	lift coefficient
C_{Lmax}	maximum lift coefficient
$G_{(n,\alpha)}$	thrust coefficient
C_s	spanwise force coefficient
C_n	normal force coefficient
C_x	force coefficient corresponding to a force acting in the x direction
C_z	force coefficient corresponding to a force acting in the z direction
c_f	coefficient of static friction
p_c	a pressure coefficient
η	efficiency
η_m	efficiency of muscular work
η_p	mechanical (Froude) efficiency
η_a	aerobic efficiency
R_e	Reynolds Number
R_d	Reynolds Number based on diameter

R_l	Reynolds Number based on length
R_h	Reynolds Number based on the height of a roughness element
$R_{l(crit)}$	critical Reynolds Number for boundary layer transition
R_x	local Reynolds Number
B	shape factor for fins
β	shape factor dependent on the distribution of cross-sectional depth of a fish
κ	a drag augmentation factor due to the influence of bodily oscillation
§	a drag augmentation factor due to the influence of the water's surface
R'	relative energy saving of burst-and-glide swimming over constant-speed swimming for a neutrally buoyant fish
R''	relative energy saving of intermittent swimming over constant-speed swimming for a negatively buoyant fish
R'''	relative energy saving of tidal-stream transport over constant speed swimming

1

Introduction to fluid dynamics

Introduction

Firstly, dimensions, units and the physics of motion are briefly reviewed. This is followed by a discussion of some basic concepts in fluid dynamics. We begin by considering the properties of fluids and the relation between fluid velocity and pressure. Consideration of inviscid (ideal) fluid flow leads us to the surprising conclusion that an obstacle placed in a flow does not experience any fluid resistance (d'Alembert's Paradox). A discussion of real (viscous) fluid behaviour based on the boundary layer concept and the phenomena of transition and separation enables us to resolve d'Alembert's Paradox. The flow of a real fluid around bluff and streamlined bodies is then discussed; discussion is centred on the drag and lift forces which they experience in a steady flow.

This chapter provides the general background information on fluid dynamics required to understand most of the discussion of fish swimming which follows. Further fluid dynamic theory is introduced later in the book. Those interested in a more detailed treatment of the topics covered in this chapter are referred to Prandtl & Tietgens (1934), Goldstein (1938), Prandtl (1952), Schlichting (1952), von Mises (1959), Hoerner (1965), Daugherty & Franzini (1977) and Tritton (1977).

Units, dimensions and a review of the physics of motion

Units and dimensions

Length (L), mass (M) and time (T) form the basis of the MLT system of dimensions. They are often referred to as primary quantities. Quantities such as area and volume are derived. In the MLT system derived quantities can be 'broken down' into their primary components. For example, area has the dimensions of length^2, or L^2. Similarly, velocity (length/time) and density (mass/volume) can be represented as LT^{-1} and ML^{-3}, respectively.

Systems of units are based on derived quantities. In this book S.I. units (Système International d'Unites) are employed. Essentially, the S.I. system

1

is an internationally agreed version of the metric system. There are seven so-called primary units; we will be concerned with three of them: the metre (m), the kilogram (kg) and the second (s). Other units are based on these. For example, the S.I. unit for force is the newton (N), $1 \text{ N} = 1 \text{ kg m}^{-1} \text{ s}^{-2}$.

For further details and an excellent source of conversion factors, see Pennycuick (1974).

Velocity and acceleration

If a particle located at a point P is displaced by an amount Δs to a new position P_1 in a time interval Δt, the average velocity of the particle can be written as $\Delta s/\Delta t$. For small intervals of time and distance we obtain the instantaneous velocity v

$$v = \frac{\mathrm{d}s}{\mathrm{d}t} \tag{1}$$

The magnitude of the velocity of the particle is referred to as its speed. The dimensions and units of speed are LT^{-1} (length × time^{-1}) and m s^{-1} (metres × seconds^{-1}), respectively.

If the instantaneous velocities of the particle are known at two points P and P_1, the average acceleration of the particle is given by $\Delta v/\Delta t$. The instantaneous acceleration a is obtained by taking small intervals of time and velocity

$$a = \frac{\mathrm{d}v}{\mathrm{d}t} \tag{2}$$

The dimensions and units of acceleration are LT^{-2} and m s^{-2}, respectively.

For constant acceleration, eqns 1 and 2 can be integrated (assuming initial conditions of $s = s_1$, $v = v_1$ and $t = 0$) to give equations which relate a, v, s and t

$$v = v_1 t + at \tag{3}$$

$$s = s_1 + v_1 t + \tfrac{1}{2}at^2 \tag{4}$$

$$v^2 = v_1^2 + 2a(s - s_1) \tag{5}$$

Force, work and power

There are three basic laws governing the motion of a particle. They were first stated by Isaac Newton in 1687. The laws are:

First law. A particle originally at rest, or moving in a straight line with a constant velocity, will continue to do so, provided the particle is not subject to an unbalanced force.

Second law. A particle acted upon by an unbalanced force F receives an acceleration a that has the same direction as the force and a magnitude that is directly proportional to the force.

Third law. For every force acting on a particle, the particle exerts a force of equal magnitude and opposite direction.

Newton's second law forms the basis of the science of dynamics. It relates the motion of a particle to the forces that act upon it. If an unbalanced force F_1 is applied to a particle, an acceleration a_1 may be measured. The force and acceleration are directly proportional and the constant of proportionality (m) may be determined from the ratio $m = F_1/a_1$. A different unbalanced force F_2 would produce an acceleration a_2; however, the ratio $F_2/a_2 = m$.

Force and acceleration are vector quantities; both have magnitude and direction. The constant of proportionality m is the mass of the particle. Mass is a scalar (magnitude only) quantity.

Newton's second law may be written as

$$F = \frac{\mathrm{d}}{\mathrm{d}t}(mv) \qquad (6)$$

or more simply as
$$F = ma \qquad (7)$$

The dimensions of force are MLT^{-2} and, as noted above, the unit is the newton. Weight is also measured in newtons. The weight of a particle is given by the product of its mass and the gravitational constant g ($W = mg$, $g = 9.81 \text{ m s}^{-2}$).

Impulse (I) is the integral of a force with respect to time. We can write

$$I = \int_{t_1}^{t_2} F \, \mathrm{d}t = \int_{v_1}^{v_2} \mathrm{d}(mv) = mv_2 - mv_1 \qquad (8)$$

The quantity mv is called momentum. Essentially, eqn 8 shows that the area under the force–time graph between times t_1 and t_2 is equal to the total increase in momentum. Impulse and momentum are both measured in newton seconds ($1 \text{ N s} = 1 \text{ kg m s}^{-1}$).

Considering F as a function of displacement, we can write

$$F = \frac{\mathrm{d}}{\mathrm{d}t}(mv) = m\frac{\mathrm{d}v}{\mathrm{d}t} = mv\frac{\mathrm{d}v}{\mathrm{d}s} \qquad (9)$$

Integrating with respect to s

$$\int_{s_1}^{s_2} F \, \mathrm{d}s = \int_{v_1}^{v_2} mv \, \mathrm{d}v = \tfrac{1}{2}mv_2^2 - \tfrac{1}{2}mv_1^2 \qquad (10)$$

Eqn 10 shows that the area under a force–displacement graph between displacements s_1 and s_2 is equal to the total increase in kinetic energy of the particle as it moves from one position to the next.

Work and energy are measured in joules (J). One joule of work is done when a force of one newton moves along its line of action one metre ($1 \text{ J} = 1 \text{ N m}$). The moment of a force (force × length of lever arm) is also measured in newton metres; however, the moment of a force is quite distinct from work and energy. A moment is a vector quantity, whereas work is a scalar.

The rate at which work is done is referred to as power. Power is a scalar quantity, expressed in $J \text{ s}^{-1}$ or Watts ($1 \text{ J s}^{-1} = 1 \text{ W} = 1 \text{ kg m}^2 \text{ s}^{-1}$). The concept of mechanical efficiency is often central to the assessment of the performance of machines and animals. Mechanical efficiency (η) is defined as the ratio of useful power output to the total power input to the system

$$\eta = \frac{\text{power output}}{\text{power input}} \tag{11}$$

Viscosity

We begin by defining a fluid as a substance that deforms continuously when subjected to a shear stress. Imagine that we place such a substance between two plates. The lower plate is fixed in position, and a force is applied to the upper plate. A shear stress develops which is equal to F/A, where A is the surface area of the upper plate. Experiments have shown that the fluid in immediate contact with a solid boundary has the same velocity as the boundary. This result is termed the 'no-slip condition'.

The fluid movement induced by moving the upper plate is illustrated diagrammatically in Fig. 1. Fluid in the area a, b, c, d flows to the new position a, b', c', d. The fluid velocity u varies from zero at the boundary of the stationary plate to U, the velocity at which the upper plate is moved. Experiments show that F is directly proportional to A and U and inversely proportional to the distance between the two plates, d. Letting $\tau = F/A$

$$\tau = \mu \frac{U}{d} \tag{12}$$

μ is a constant of proportionality called the viscosity of the fluid.

A more general expression of eqn 12 is

$$\tau = \frac{\mathrm{d}u}{\mathrm{d}y} \mu \tag{13}$$

where the velocity gradient $\mathrm{d}u/\mathrm{d}y$ can be thought of as the rate at which one layer of the fluid moves relative to the next. Eqn 13 is known as Newton's Law of Viscosity. The dimensions of τ, u, and y are FL^{-2}, LT^{-1} and L, respectively.

Solving eqn 13 for μ gives

$$\mu = \frac{\tau}{\mathrm{d}u/\mathrm{d}y} \tag{14}$$

and the dimensions of μ as $ML^{-2}\,T^{-2}$. The S.I. unit of viscosity is the pascal second (1 Pa s $= 1$ N s m$^{-2} = 1$ kg m^{-1} s^{-1}).

Viscosity is not to be confused with the kinematic viscosity, ν

$$\nu = \mu/\rho \tag{15}$$

where ρ is the fluid density. The dimensions and S.I. units of kinematic viscosity are L^2T^{-1} and m^2 s^{-1}, respectively. Kinematic viscosity has many applications, some of which will be discussed later.

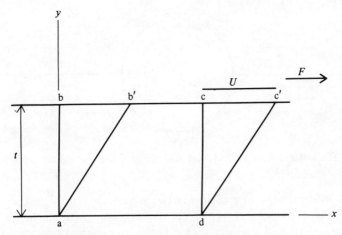

Fig. 1 Fluid deformation as a result of the application of a constant shear force.

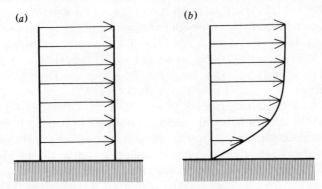

Fig. 2 The velocity profile of an ideal fluid (*a*) and of a real fluid (*b*) at a solid boundary.

Ideal and real fluids

For the purposes of mathematical analysis, hydrodynamacists assume that a fluid is inviscid. If the fluid can also be considered to be incompressible, it is termed an ideal or perfect fluid. Water may be treated as an ideal fluid. For large objects moving rapidly through water, the inertial forces far exceed the viscous forces and the results obtained from ideal fluid theory correspond well with the experimental observations on real fluids.

With zero viscosity however, no tangential stresses can occur within the fluid and therefore ideal fluids 'slip' with respect to solid boundaries (Fig. 2*a*). This theoretical result leads to the conclusion that a body moving at a uniform speed will not be subject to a retarding drag force!

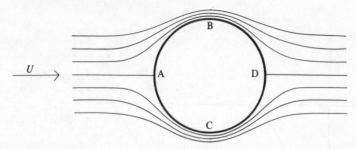

Fig. 3   The inviscid flow around a circular section. See the text for a discussion.

Fig. 3 illustrates the case of steady two-dimensional flow past a circular cylinder. Fluid particles travel in parallel straight lines which are called streamlines. The fluid particles move at the undisturbed free stream fluid velocity U. At stations B and C the streamlines are crowded together. The mass flow per unit time across any section between the same two streamlines is constant, and therefore, a crowding of the streamlines denotes an increase in the fluid velocity. At points A and D the fluid velocity is zero (such points are called stagnation points). So, as we pass along the cylinder from A to B or C the fluid velocity increases. In passing from B or C to D the fluid velocity falls to zero.

The increase in velocity from A to B or C is accompanied by a pressure gradient in the upstream direction and the decrease in velocity from B or C to D is associated with a pressure gradient in the opposite direction. In passing from A to B or C the fluid acquires enough momentum (in the absence of friction) to oppose the adverse pressure gradient from B or C to D. The pressures on the front half of the cross-section (B, A, C) are exactly balanced by those on the rear half (B, C, D), so there can be no net drag force. Water, like all other real fluids, has viscosity and objects which move through it are subject to a retarding drag force. The discrepancy in behaviour between real and ideal fluids is known as d'Alembert's Paradox.

Continuity and Bernoulli's theorem

The stream tube is a useful concept in fluid dynamics. We can loosely describe a stream tube as being made up of all of the streamlines which pass through any imaginary, closed curve drawn in a flow field. There is no flow through the walls of the tube, only along its length. A stream filament is a stream tube of infinitesimal cross-section.

A stream filament is illustrated in Fig. 4. Consider the steady flow of an inviscid fluid through the filament. If ρ_1, U_1 and ρ_2, U_2 are the density and

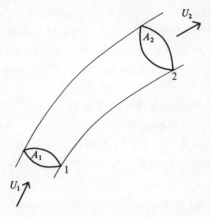

Fig. 4   Flow through a stream filament. See the text for an explanation.

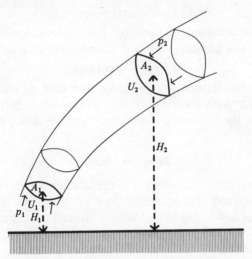

Fig. 5   Flow through a stream tube.

velocity at the cross-sections A_1 and A_2, respectively, then the rate of mass flow into the region bounded by A_1 must be equal to that out of A_2

$$\rho_1 U_1 A_1 = \rho_2 U_2 A_2 \tag{16}$$

Eqn 16 is known as the continuity equation. For incompressible flow, ρ is a constant and eqn 16 reduces to

$$UA = \text{constant} \tag{17}$$

In Fig. 5 we introduce the terms p_1, H_1 and p_2, H_2 to represent the pressure and height above a fixed reference level at A_1 and A_2, respectively. The flow is incompressible, and so $\rho_1 = \rho_2 = \rho$. By continuity $A_1 U_1 = A_2 U_2 = AU$ and so the work done by the pressure forces at A_1 and A_2 is

$$W_T = p_1 A_1 U_1 \, dt - p_2 A_2 U_2 \, dt = AU(p_1 - p_2) \, dt \tag{18}$$

where W_T is the total work done on the fluid. This term is made up of the gain in kinetic energy W_K

$$W_K = \tfrac{1}{2}\rho_2 U_2 A_2 \, dt \, U_2{}^2 - \tfrac{1}{2}\rho_1 U_1 A_1 \, dt \, U_1{}^2 = \tfrac{1}{2}\rho(UA_2{}^2 - U_1{}^2) \, dt \tag{19}$$

plus the gain in potential energy W_P

$$W_P = g\rho_2 U_2 A_2 \, dt \, H_2 - g\rho_1 U_1 A_1 \, dt \, H_1 = g\rho UA(H_2 - H_1) \, dt \tag{20}$$

so $AU(p_1 - p_2) \, dt = \tfrac{1}{2}\rho UA(U_2{}^2 - U_1{}^2) \, dt + g\rho UA(H_2 - H_1) \, dt$

$$p_1 - p_2 = \tfrac{1}{2}\rho U_2{}^2 - \tfrac{1}{2}\rho U_1{}^2 + g\rho H_2 - g\rho H_1$$

$$p_1 + \tfrac{1}{2}\rho U_1{}^2 + g\rho H_1 = p_2 + \tfrac{1}{2}\rho U_2{}^2 + g\rho H_2 \tag{21}$$

In the limiting case of a single streamline we have

$$p + \tfrac{1}{2}\rho U^2 + g\rho H = \text{constant} \tag{22}$$

Eqn 22 is known as Bernoulli's equation. For most purposes the value of $g\rho H$ can be treated as a constant, and eqn 22 can be written as

$$p + \tfrac{1}{2}\rho U^2 = \text{constant} \tag{23}$$

It follows from eqn 23 that, for any given flow field, an increase in velocity is accompanied by a decrease in pressure, and vice-versa. Bernoulli's equation also shows that the pressure is low at points in the flow where the velocity is high.

Reynolds Number

In a series of experiments conducted in the late nineteenth century, Osborne Reynolds investigated the nature of flow in tubes. He introduced a dye filament into the flow of liquid as it flowed through a narrow tube. Reynolds observed that on some occasions the dye passed through the tube as an unbroken filament and that on other occasions the filament broke up. The two situations correspond to laminar and turbulent flow, respectively.

Reynolds noted that the nature of the flow (whether it is laminar or turbulent) is dependent on the viscosity, density, fluid velocity and the tube diameter, and that the influence of all of these factors can be represented by a single non-dimensional index, which we now call the Reynolds Number, R_d.

$$R_d = \frac{\rho U d}{\mu} \tag{24}$$

where d is the tube diameter. For general flow conditions the Reynolds Number is written as

$$R_l = \frac{\rho l U}{\mu} \tag{25}$$

where l is any specified characteristic length. In the case of water the values of ρ and μ are constant at any given temperature and so the kinematic viscosity is often substituted into the equation

$$R_1 = \frac{lU}{\nu} \tag{26}$$

Essentially, eqn 26 expresses the relative importance of the inertial and viscous forces which act on a submerged body. When R_1 is small, viscous forces dominate over inertial effects and the flow is laminar. At high values of R_1 viscous forces are relatively small and the flow is turbulent.

The flows around geometrically similar bodies are said to be dynamically similar if their Reynolds Numbers are the same. The concept of dynamical similarity in flow conditions is important as it forms the basis of model testing.

Boundary layer theory

In order to account for drag the viscous effects of fluids have to be considered. Viscosity gives rise to the no-slip condition in a real fluid. There is no relative velocity between the surface of a submerged body and the particles of fluid in contact with it. A short distance from the surface of the body, however, the fluid velocity approximates that of the free stream. The region over which this velocity increase occurs is called the boundary layer. Viscous stresses are large in the boundary layer due to the large velocity gradients that occur there. One way of representing the change in velocity in the boundary layer is illustrated in Fig. 2b.

Boundary layer theory can be employed to account for drag in two ways. Firstly, the boundary layer gives rise to a modified distribution in the pressure over a body and a net resultant pressure in the stream direction. This type of drag is called form or pressure drag. Secondly, when integrated over the entire surface of the body the tangential stress in the boundary layer also gives a net resultant in the stream direction. This type of drag is termed skin friction drag. Taken together, the form and frictional drag of a body are often referred to as the profile drag. Further discussion of drag forces and their classification will be given later in this chapter.

Experiments have shown that there are two basic types of boundary layer flow, laminar and turbulent. Laminar boundary layers are characterized by a flow which is smooth and steady. The laminar boundary layer is very thin, and therefore the form drag is small. Due to the moderate velocity gradients which occur across the laminar boundary layer, frictional forces are also relatively small.

In the turbulent boundary layer, flow is unsteady and eddying. Due to this it is necessary to report mean velocities in the boundary layer rather than instantaneous values. Over a given time interval the velocity distribution will be the same as that over another such interval, and so the concept of the stable

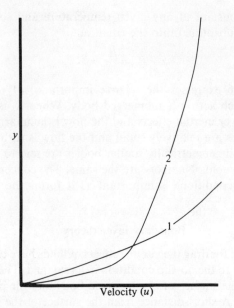

Fig. 6   The velocity profile at a solid boundary for the case of a laminar (1) and a turbulent (2) boundary layer.

velocity profile is still valid and useful. The eddying nature of the flow gives rise to a boundary layer that is thicker than the laminar case (other things being equal) and because of this the resulting form drag is higher.

Eddies in the turbulent boundary layer facilitate energy exchange between given stations. A consequence of this is that the velocity near to the surface of the body will be higher than in the case of a laminar boundary layer, where no such energy exchange occurs. This implies that the velocity gradient near the body will be relatively higher, and therefore so will the frictional drag. A diagrammatic comparison of the velocity profile of the laminar and turbulent boundary layers is given in Fig. 6.

For flat plates and streamlined bodies under steady flow conditions the boundary layer can be expected to be laminar up to Reynolds Numbers of about 5×10^5 and turbulent when R_1 is greater than about 5×10^6. At intermediate values of R_1 the flow is characterized by mixed laminar and turbulent regions. This type of flow is referred to as transitional. The value of R_1 at which transition occurs is called the critical Reynolds Number.

Boundary layer thickness is usually expressed as the velocity thickness δ, which is equal to the distance from the surface of the body to which the boundary layer is attached to the position where the velocity in the boundary layer differs by 1% from the free-stream velocity.

In flow along a flat plate orientated parallel to the stream, the thickness of the boundary layer is dependent not on the overall length of the plate, but on the distance x of the section being considered from the forward edge of the plate. The thickness of the boundary layer grows parabolically and varies as $(vx/U)^{\frac{1}{2}}$. The shear stress at the surface of the plate is inversely proportional to the boundary layer thickness (i.e. it varies as $x^{-\frac{1}{2}}$).

For laminar boundary layers the velocity thickness is approximately related to the Reynolds Number by

$$(\delta/x) \simeq 5.0(R_1^{-0.5}) \tag{27}$$

Similarly, for the case of a turbulent boundary layer

$$(\delta/x) \simeq 0.37(R_1^{-0.2}) \tag{28}$$

In the case of a transitional boundary layer the velocity thickness is given by

$$\delta = 5.0(R_{1(\text{crit})}^{-0.5}) \tag{29}$$

where $R_{1(\text{crit})}$ denotes the critical Reynolds Number.

Transition and separation

Initially, the flow over a solid body such as a flat plate or aerofoil orientated parallel to the stream is laminar. Moving downstream however, a region is reached where the laminar boundary layer breaks down and becomes turbulent. The region in which this change occurs (transition region) is characterized by a sudden increase in the thickness of the boundary layer.

The most important factor determining the onset of transition is the local Reynolds Number, R_x

$$R_x = \frac{\rho U x}{\mu} \tag{30}$$

where x is the distance from the leading edge of the body. At low values of R_x (less than about 5×10^5 for a flat plate orientated parallel to the flow) viscous forces are sufficiently large to damp out disturbances in the boundary layer. At higher R_x, disturbances in the boundary layer are amplified by the dominant inertial forces. This results in transition and the development of a turbulent boundary layer.

In the absence of a pressure gradient the boundary layer over a flat plate will continue to grow in the downstream direction, regardless of the length of the plate. For bodies which are characterized by a strong adverse pressure gradient (pressure increasing rapidly in the stream direction) this is not the case. In such cases, fluid particles that are close to the surface of the body are retarded by the pressure forces. If the adverse pressure gradient is large enough, the velocity near to the surface will fall and eventually be reversed in direction (Fig. 7). If this occurs the streamlines are lifted up from the surface and large eddies are formed. Both laminar and turbulent boundary layers can separate (Fig. 8). Other things being equal, turbulent boundary layers separate further back than laminar ones and produce a narrower wake (Fig. 8). Turbulent boundary layers tend to separate less readily than laminar

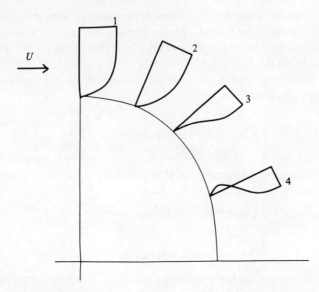

Fig. 7   Flow reversal over a spherical body. Flow profiles are shown at four different positions on the sphere.

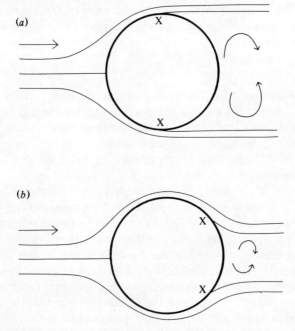

Fig. 8   Flow separation around a sphere, for the case of a laminar (*a*) and turbulent (*b*) boundary layer.

ones. The laminar boundary layer requires only a relatively short region of adverse pressure to cause separation. Once separation has occurred the adverse pressure gradient is destroyed and a sudden increase in form drag occurs. If the body was producing a lift force prior to separation, this force is markedly reduced. Separation is often associated with unsteadiness in the flow and this can give rise to buffeting of the body.

Free-stream turbulence (not to be confused with boundary layer turbulence) is a common cause of transition and separation. Essentially, the inertial forces generated by free-stream turbulence can be viewed as leading to an increase in the local Reynolds Number beyond that at which transition occurs. The intensity of turbulence, I, is measured by relating the root mean square velocity of the fluid in the x, y and z planes ($\overline{u^2}$, $\overline{v^2}$ and $\overline{w^2}$, respectively) to the free-stream velocity. For isotropic flow ($\overline{u^2} = \overline{v^2} = \overline{w^2}$) the magnitude of the velocity fluctuations is the same in all three axes and

$$I = \frac{(\overline{u^2})^{\frac{1}{2}}}{U}$$ (31)

For values of I greater than about 0.02–0.03 the impact of disturbances in the boundary layer is amplified and transition is induced.

Transition and separation can also be caused by surface roughness. In order to be effective in causing transition, the roughness elements must protrude through the boundary layer. The inertial effects produced by roughness elements which lie within the boundary layer are quickly damped out by viscous forces. The likely influence of roughness elements on the flow over a body can be determined from a Reynolds Number, R_h

$$R_h = \frac{Uh}{\nu}$$ (32)

where h is the height of the element. Transition will be caused when R_h is greater than about 900 for a single roughness element and about 120 for distributed elements.

Real fluid flow past a bluff symmetrical body

Earlier in this chapter the flow around a circular section in an ideal fluid was considered. We noted that there can be no drag (d'Alembert's Paradox), as the pressures on the front half of the section are exactly balanced by those on the rear half. The flow around the same section for the case of a real fluid is illustrated in Fig. 9. The flow is characterized by a boundary layer which is subject to pressure gradients. From A to B or C the pressure gradient is in the stream direction, and is therefore favourable. Passing from B or C to D however, the gradient is in the opposite direction. This gives rise to flow reversal in the boundary layer, separation and the formation of an extensive wake.

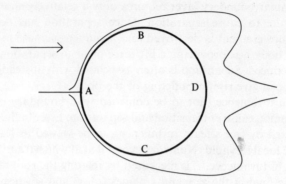

Fig. 9   The flow pattern of a real fluid around a sphere. See the text for a discussion.

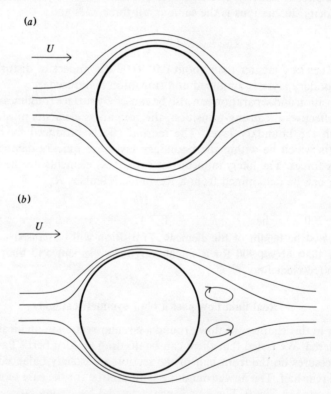

Fig. 10   Low Reynolds Number flow around a circular cylinder (a, $R_d = 1$; b, $R_d = 40$). Modified from Tritton (1977).

Separation of the flow and the formation of a wake prevents a rise in pressure towards the rear of the body. Consequently, there is a net downstream pressure, and therefore a drag force (form drag) in addition to the viscous drag on the body due to the tangential shear forces generated in the boundary layer.

The flow pattern is dependent on the Reynolds Number. At low Reynolds Numbers ($R_d < 1$) viscous forces dominate over the pressure terms which can be considered as being negligible. Due to this the flow around the section is symmetrical upstream and downstream (Fig. 10*a*). At higher R_d this symmetry of the flow is lost. When R_d is greater than about 4, two 'attached eddies' are seen behind the section (Fig. 10*b*); as R_d increases these eddies increase in size. At R_d greater than about 40 the flow becomes unsteady. This instability gives rise to a flow pattern known as the Kármán vortex street (after von Kármán who first described it). The vortex street consists of two rows of rapidly rotating fluid vortices. The vortices are shed alternately from the body and this gives rise to a staggered arrangement. All of the vortices on one side of the street rotate in the same direction, which is the opposite to that of those on the other side.

The frequency of vortex shedding is given by the empirical formula

$$\frac{nd}{U} = 0.198\left(1 - \frac{19.7}{R_d}\right) \tag{33}$$

for $250 < R_d < 2 \times 10^5$, where d is the diameter of the section and n is the frequency of vortex shedding. The non-dimensional parameter nd/U is called the Strouhal Number. The Strouhal Number varies slowly with R_d and values of about 0.2 are typical for R_d of $10^3 - 5 \times 10^5$. At a Reynolds Number of about 200 a further change in the flow occurs. The wake becomes less well organized and turbulent. Despite turbulence however, periodic vortex shedding continues up to the highest Reynolds Numbers that have been investigated ($R_d > 10^7$).

Streamlining

At higher Reynolds Numbers ($> 10^3$) the drag of a bluff body is mainly due to the underpressure that develops in the wake. This can be markedly reduced by producing a profile in which boundary layer separation is delayed till a point near to the rear of the section. Such profiles are said to be streamlined. Streamlined profiles are characterized by a slowly tapering tail and a ratio of maximum length to maximum diameter (Fineness Ratio) of between 2 and 6. For least resistance to motion, a streamlined body of given volume should have a Fineness Ratio of about 4.5. It can be seen from Fig. 11 that this is not critical however.

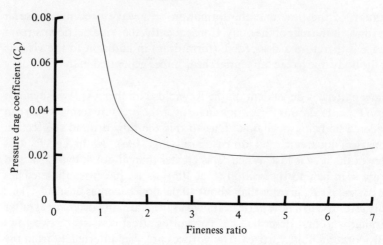

Fig. 11   Influence of fineness ratio on the drag of a streamlined body. Modified from von Mises (1959).

Fluid dynamic drag and lift

Friction drag

Consider a two-dimensional plate of length l (in the stream direction) and breadth b, moving in its own plane. It is reasonable to assume that all of the drag is due to shear stress in the boundary layer over the plate. By employing boundary layer theory it is possible to calculate the amount of friction drag developed on both sides of the plate. It can be shown that the friction drag D_f is given by

$$D_f = \tfrac{1}{2}\rho U^2 S_w\, C_f \tag{34}$$

where S_w is the total wetted surface area of the plate ($= 2bl$) and C_f is a frictional drag coefficient. C_f is given by

$$C_f = \frac{2D_f}{\rho U^2 S_w} \tag{35}$$

The value of C_f depends on the boundary layer flow regime, which in turn depends on the Reynolds Number. For a laminar boundary layer and values of R_1 up to about $250\,000$–$300\,000$

$$C_{f(\text{lam})} = \frac{2D_f}{\rho U^2 S_w} = 1.33\sqrt{\frac{\mu}{\rho U L}} = \frac{1.33}{\sqrt{R_1}} \tag{36}$$

For the case of a turbulent boundary layer, C_f is given by

$$C_{f(\text{turb})} = \frac{2D_f}{\rho U^2 S_w} = 0.072\sqrt[5]{\frac{\mu}{\rho U L}} = \frac{0.072}{\sqrt[5]{R_1}} \tag{37}$$

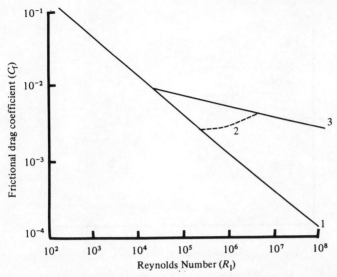

Fig. 12 Relationship between the friction drag coefficient and Reynolds Number for a flat plate with a laminar (1), transitional (2) or turbulent (3) boundary layer.

In the transitional region of Reynolds Numbers the frictional drag coefficient can be calculated from

$$C_{f(tran)} = \frac{0.072}{\sqrt[5]{R_1}} - \frac{1700}{R_1}$$ (38)

Values of $C_{f(lam)}$, $C_{f(turb)}$ and $C_{f(tran)}$ are plotted against R_1 in Fig. 12.

The arguments outlined above are simplifications. For a plate of sufficient length the boundary layer at the front will be laminar. At some distance back from the leading edge, transition will occur and the boundary layer will become turbulent. However, the formulae given above can be employed to give reasonable estimates of the friction drag coefficient of flat plates and streamlined bodies.

Pressure drag

We have already discussed the origin of pressure drag. For bluff bodies the pressure drag coefficient C_p and the Reynolds Number are essentially independent over a large Reynolds Number range. At a certain critical Reynolds Number range (the transition region) however, the value of C_p falls dramatically. For Reynolds Numbers greater than those of the transition range values of C_p are relatively constant again.

The changes in the value of C_p with Reynolds Number reflect the existence of two distinct boundary layer flow regimes, separated by a transitional region. At moderate Reynolds Numbers the boundary layer is laminar. Beyond the transitional region boundary layer flow is turbulent. The pressure

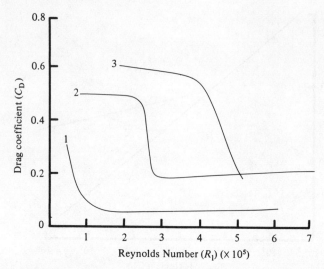

Fig. 13   Drag of a sphere (2), ellipsoid with a ratio of major to minor axis of 1:1.8 placed with its major axis parallel to the flow (1) and an ellipsoid with a ratio of major to minor axis of 1:0.75 placed with its major axis normal to the flow. Modified from von Mises (1959).

drag resulting from separation of a laminar boundary layer is higher than that for a turbulent one. This is due to the fact that the point of separation is further forward in the case of the laminar boundary layer and in consequence a broader wake develops (see Fig. 8).

For the case of a sphere the range of moderate R_d extends from about 20 000 to about 200 000 and C_p is approximately 0.5. At values of R_d greater than about 300 000, C_p is about 0.2. The relation between R_d and C_p is illustrated in Fig. 13 for the cases of the sphere and ellipsoid.

The pressure drag of streamlined bodies is small relative to the skin friction drag and it is usual for the pressure drag coefficient to be calculated as a fraction of the frictional drag coefficient via

$$C_{Do} = (C_f + C_p) = C_f(1 + 1.5(d/l)^{\frac{3}{2}} + 7(d/l)^3) \qquad (39)$$

where d is the maximum width of the section and C_{Do} is the sum of the skin friction and pressure drag coefficients (profile drag coefficient).

Interference drag

If an appendage is added to the surface of a well streamlined section, the resultant drag of the system is much larger than the sum of the drag on the appendage plus that of the streamlined body when tested independently. The additional drag term which arises due to the interaction of the flow around the two bodies is called the interference drag.

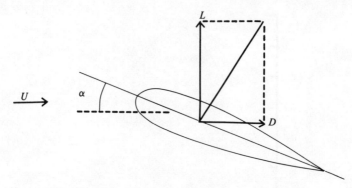

Fig. 14 Lift and drag forces acting on an aerofoil.

The magnitude of the additional drag term depends on the position at which the appendage is placed. The interference drag reaches a maximum when the source of disturbance is placed close to the point of maximum cross-section (shoulder) of the main body. When bluff appendages (plates normal to the flow, etc.) are placed in this position an increase in the drag of the main body of up to five times can occur.

Lift

In addition to drag, a body moving through a real fluid may experience a lift force L which acts normal to the drag force

$$L = \tfrac{1}{2}\rho U^2 S_w C_L \tag{40}$$

where C_L is a lift coefficient, which is given by

$$C_L = \frac{2L}{\rho U^2 S_w} \tag{41}$$

The lift and drag components acting on a streamlined body (in this case an aerofoil) are illustrated in Fig. 14.

The value of C_L varies with the angle of incidence (sometimes referred to as the angle of attack) of the section (α, in Fig. 14). Fig. 15 shows how the value of C_L for a typical aerofoil section varies with the angle of incidence. In this case, zero lift is produced at a small negative angle of incidence. As α increases the lift curve becomes linear. This part of the lift curve can be represented by an equation of the form

$$C_L = a(\alpha - \alpha_0) \tag{42}$$

where a is a constant. The lift curve slope is given by $dC_L/d\alpha = a$. At a certain value of α separation begins to occur and the slope of the curve falls off. When the flow has fully separated from the rear surface of the section the aerofoil is said to be stalled. The stalling angle corresponds to the maximum value of C_L, which is denoted by C_{Lmax}. For an aerofoil without high-lift devices a value of $C_{Lmax} = 1.2$–1.4 at $\alpha \simeq 15°$ would be typical.

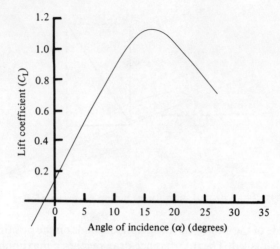

Fig. 15   The variation of the lift coefficient with angle of incidence for an aerofoil. Data from von Mises (1959).

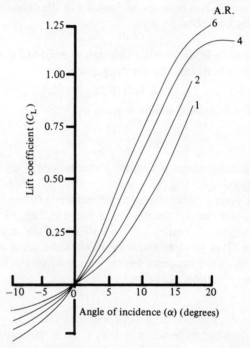

Fig. 16   The influence of aspect ratio on the lift coefficient. Modified from von Mises (1959).

The value of C_{Lmax} depends on: (1) aspect ratio; (2) Reynolds Number; (3) camber; (4) thickness; and (5) nose radius.

The aspect ratio of an aerofoil (A.R.) is defined as the span (s) divided by the mean geometric chord $(\bar{c})$. If A denotes the plan area of the body, we can write

$$\text{A.R.} = \frac{b}{\bar{c}} = \frac{b^2}{b\bar{c}} = \frac{b^2}{A} \tag{43}$$

The influence of aspect ratio on the lift slope curve is illustrated in Fig. 16.

At higher Reynolds Numbers the boundary layer flow over the surface of an aerofoil is turbulent. The turbulent boundary layer is less prone to separation than is the laminar boundary layer, and therefore higher values of C_{Lmax} can be attained at higher Reynolds Numbers.

Aerofoils with an arched centre line are said to be cambered. Cambered sections produce higher lift at all angles of incidence, and although the stall occurs slightly earlier, values of C_{Lmax} are generally higher than those for uncambered sections. The value of the maximum thickness of an aerofoil divided by the chord is called the thickness ratio (t/c). Values of C_{Lmax} increase from low values of t/c up to values of t/c of about 12–14%. At higher values of t/c, C_{Lmax} falls off. A sharp nose (i.e. leading edge of small radius of curvature) is usually associated with a low thickness ratio. If the leading edge of a section is too sharp, early separation and a leading edge stall can result.

Profile drag and the lift to drag ratio

The sum of the friction drag and pressure drag forces that act on a streamlined section or aerofoil is referred to as the profile drag. At low angles of incidence most of the profile drag is due to skin friction. At higher angles of incidence the wake behind the section widens and the pressure drag term becomes dominant.

One measure of the efficiency of an aerofoil is the lift to drag ratio L/D. We can write

$$\frac{L}{D} = \frac{\frac{1}{2}\rho U^2 S_{\text{w}} C_{\text{L}}}{\frac{1}{2}\rho U^2 S_{\text{w}} C_{\text{Do}}} = \frac{C_{\text{L}}}{C_{\text{Do}}} \tag{44}$$

where C_{Do} denotes the profile drag coefficient. The values of $C_{\text{L}}/C_{\text{Do}}$ and C_{Do} are plotted against α in Fig. 17 which shows that: (1) $C_{\text{L}}/C_{\text{Do}}$ increases rapidly up to $\alpha \simeq 10°$ and C_{Do} remains relatively constant over the same interval; and (2) for angles of incidence greater than about $10°$, $C_{\text{L}}/C_{\text{Do}}$ falls off and C_{Do} rises rapidly.

At low angles of incidence the value of C_{L} increases linearly with increasing α. The wake behind the section is narrow and skin friction forces dominate over pressure drag. As α increases, C_{L} approaches C_{Lmax} and falls off. At the same time the wake widens causing an increase in pressure drag and a sharp rise in C_{Do}. For good efficiency the value of $C_{\text{L}}/C_{\text{Do}}$ should be large. The changes discussed above are commonly represented by a polar diagram in which C_{L} is plotted against C_{Do}; an example is given in Fig. 18.

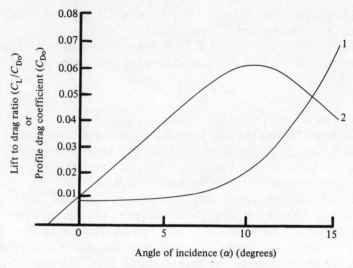

Fig. 17 The variation of the lift to drag ratio (1) and the profile drag coefficient (2) with angle of incidence for an aerofoil.

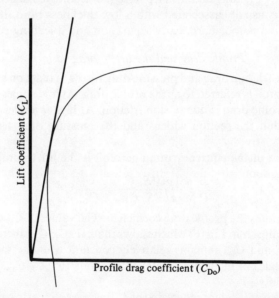

Fig. 18 A polar plot of the lift and drag coefficients of an aerofoil. The tangent to the curve gives the position at which the maximum lift to drag ratio is obtained.

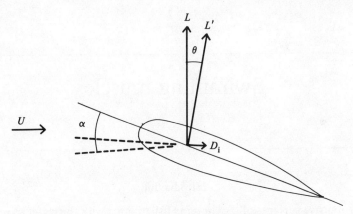

Fig. 19   Induced drag. This figure is explained in the text.

Induced drag

The lift produced by an aerofoil has its reaction in the downward momentum which it imparts to the fluid as it flows over the section. The lift of the section is equal to the rate of transport of downward momentum of the fluid. A drag force is associated with this downward deflection of fluid which is termed the induced drag (Fig. 19). The angle at which the fluid is deflected downward is called the downwash angle and is given by

$$\theta_{\mathrm{d}} = \frac{C_{\mathrm{L}}}{\pi \mathrm{A.R.}} \tag{45}$$

This implies that the effective incidence of the aerofoil is reduced by θ_{d} (see Fig. 19). The effective lift force is tilted through the same angle and this gives rise to a component in the free stream direction which is termed the induced drag D_{i}. The component of the effective lift which is normal to the free stream is given by $L = L' \cos \theta_{\mathrm{d}}$, where L' is the effective lift force. The induced drag is therefore equal to $L' \sin \theta_{\mathrm{d}}$. So, $D_{\mathrm{i}} = L' \tan \theta_{\mathrm{d}}$ and as θ_{d} is small $D_{\mathrm{i}} \simeq L\theta_{\mathrm{d}}$. At the aerofoil $\theta_{\mathrm{d}} = C_{\mathrm{L}}/\pi \mathrm{A.R.}$ and therefore

$$C_{\mathrm{Di}} = \frac{C_{\mathrm{L}}^{2}}{\pi \mathrm{A.R.}} \tag{46}$$

Eqn 46 shows that C_{Di} decreases with increasing aspect ratio and that there can be no induced drag force at zero lift.

2

Swimming muscle

Introduction

The propulsive surfaces of a swimming fish interact with the water to produce a forward thrust force which is generated by the contraction of the propulsive musculature. This chapter is concerned with the arrangement and mechanical properties of these muscles.

Firstly, the form of the force versus velocity (P–V) curve of vertebrate striated muscle is discussed in relation to power output and mechanical efficiency. The influence of temperature on power output and efficiency is considered. It is shown that the power output of unacclimated muscles in isolated preparations increases with temperature, although efficiency is not greatly affected. The validity of extending these findings to fish from different climatic regions is discussed. There is some evidence to suggest that the power output per unit mass of swimming muscle of antarctic and tropical fish may not be very different.

The propulsive musculature of any given fish is not homogenous. There are several distinct fibre types which are designed to operate at different swimming speeds. The characteristics of the so-called red and white muscle fibre types are discussed in terms of activity level. The relative proportions of red and white muscle fibres in the propulsive musculature of different species of fish is related to their mode of life.

Finally, the functional significance of the unique muscle fibre geometry found in fishes is discussed. The red fibres are superficial and run longitudinally, but the deeper white fibres are arranged in a helical way. It is shown that this arrangement of fibres is advantageous during bending of the body. Further discussion of muscle mechanics can be found in Alexander (1973), Weis-Fogh & Alexander (1977) and White (1977). Readers interested in the biochemical basis of muscular contraction and the biochemical adaptations of fish muscle are referred to reviews by Goldspink (1977) and Driedzic & Hochachka (1978).

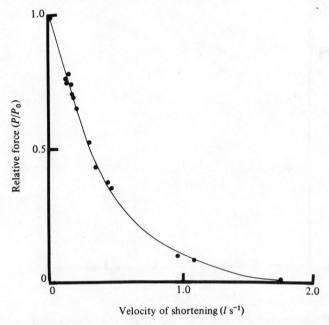

Fig. 20   Force–velocity curve for red muscle fibres from the adductor operculi muscles of *Tilapia mossambica*. From Flitney & Johnston (1979).

Muscle mechanics

Power for swimming is produced by muscles which convert metabolic fuel (carbohydrates, fats) into mechanical work. The power that a muscle produces depends upon the velocity (V) at which it contracts (shortens) and the force (P) that it exerts. The relationship between the two is described by Hill's equation (Hill, 1939)

$$V(P+a) = b(P_0 - P) \qquad (47)$$

where P_0 is the force exerted by the muscle in isometric contraction (i.e. the value of P when $V = 0$) and a and b are constants, with the units of force and velocity, respectively. In Fig. 20, relative force (P/P_0) is plotted against velocity of shortening for a striated muscle fibre from the branchial musculature of a fish. The form of the curve is typical of many muscles.

It can be shown (by differentiating eqn 47) that the greatest power output is obtained at a value of P which is given by

$$(a^2 + aP_0)^{-\frac{1}{2}} - a \qquad (48)$$

For a large variety of vertebrate striated muscles the ratio a/P_0 takes values between 0.15 and 0.25, corresponding to values of P for maximum power of

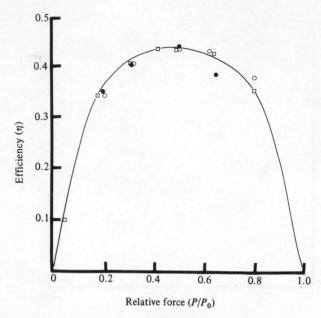

Fig. 21   Relation between efficiency and load in developing power during isotonic shortening. From Hill (1964).

$0.27P_0$ and $0.31P_0$, respectively (Hill, 1939; Alexander, 1973). When PV (the power output) is a maximum, $P/P_0 = V/V_0$ (where V_0 is the speed of shortening with zero load), and so maximum power is developed at about $P = 0.3P_0$, $V = 0.3V_0$.

Like power output the efficiency (mechanical work done/total energy used) varies with load. From studies on maintained isotonic contractions on frog sartorii, Hill (1939, 1964) obtained values for maximum mechanical efficiency of about 0.45 at $P/P_0 = 0.5$ (Fig. 21). When a muscle shortens and relaxes isotonically the work done on the load during shortening appears as heat during relaxation. In experiments on maintained isotonic contractions the muscle is not allowed to relax, and so this heat is not measured, and therefore the calculated efficiencies are about twice those to be expected of repeated contractions.

Hill (1950) summarizes the relation between velocity of shortening, force, mechanical power developed, total work used and mechanical efficiency for a typical striated muscle. Fig. 22 shows that the value of maximum efficiency of a little over 20% occurs at $0.2V_0$ with a load close to $0.5P_0$. The efficiency versus load curve is rather flat (see Figs 21 and 22) and only falls below 90% of its maximum value for loads $< 0.24P_0$ and $> 0.7P_0$. This means that muscles can work at maximum power with near maximum efficiency. At

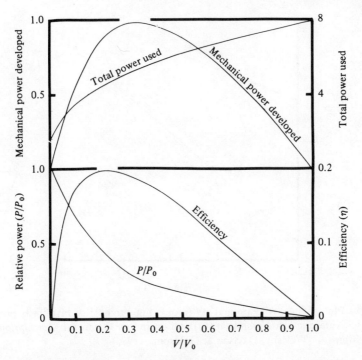

Fig. 22　Relation between various mechanical properties of muscle and speed of shortening. From Hill (1950).

speeds of shortening which are very high or low, both power and efficiency fall off rapidly.

From eqn 47, power (PV) can be calculated from

$$PV = bP\frac{P_0 + a}{P + a} - 1 \qquad (49)$$

(Hill, 1939). Substituting $a = 0.25P_0$ and $P = 0.3P_0$ into eqn 49 and rearranging, Machin (in Bainbridge, 1961) derived the expression

$$PV = 0.38bP_0 \qquad (50)$$

Using Hill's experimental data, Machin has calculated the specific power output (power output per unit mass of muscle) for frog sartorii during repeated contractions. Calculated values are temperature-dependent, ranging from about 40 W kg^{-1} at 0 °C to 170 W kg^{-1} at 30 °C. This is because the value of b increases by about 2.05 times for a 10 °C increase in temperature.

On the basis of Machin's calculations, Bainbridge (1961) suggests specific power output values of 20 W kg^{-1} and 80 W kg^{-1} for sustained and burst swimming, respectively, in cold-water fish. Values of 40 W kg^{-1} and 170 W kg^{-1} are suggested for fish from tropical waters.

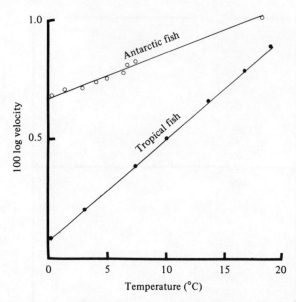

Fig. 23   Plot of $\log_{10}$ ATPase activity (mol Pi mg^{-1} min^{-1}) of myofibrils prepared from the Antarctic fish *Notothenia* ($\bigcirc$) and an Indian Ocean fish *Amphiprion* ($\bullet$). From Johnston, Walesby, Davison & Goldspink (1975).

However, although specific power output may vary with temperature within given species of fish, it seems likely that it may not be very different between different species that are adapted to function at different temperatures. There is some evidence to support this view. Muscle fibres from the Antarctic fish *Notothenia* are characterized by certain biochemical adaptations (e.g. high ATPase activity, see Fig. 23) which enable the muscles to shorten at reasonable velocities at low temperatures (Johnston, Walesby, Davison & Goldspink, 1975). Carp can acclimate to temperatures in the range 1 °C to 20 °C. They are able to synthesize low and high temperature forms of myosin (Johnston *et al.*, 1975, see Fig. 24).

It is of interest to consider whether or not muscle efficiency is influenced by temperature. We can write

$$\eta_m = \frac{PV}{PV+(q+jV)} \tag{51}$$

where η_m is the mechanical efficiency of the muscle, q is the rate of production of maintenance heat and jV is the rate at which shortening heat is produced. For a/P_0 close to 0.25, $q \simeq ab$ and $j = 0.16P_0 + 0.18P$ (Hill, 1964). Substituting these values and eqn 50 into eqn 51, we have

$$\eta_m = \frac{0.38bP_0}{ab+(0.16P_0+0.18P)V+0.38bP_0} \tag{52}$$

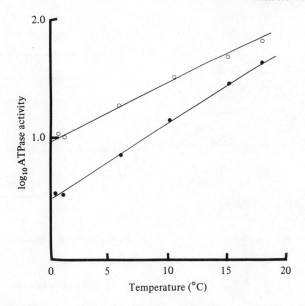

Fig. 24   Plot of $\log_{10}$ ATPase activity (mol Pi mg^{-1} min^{-1}) of myofibrils prepared from the white myotomal muscle of goldfish acclimated to 1 °C (○) and 26 °C (●). From Johnston *et al.* (1975).

For $P = 0.3P_0$, we have

$$\eta_{\mathrm{m}} = \frac{0.38b}{0.21V + 0.63b} \tag{53}$$

Substituting experimental data from Hill (1939) into eqn 53 ($a/P_0 = 0.23$, $b = 0.33$ at 0 °C, temperature coefficient = 2.05) gives values of η_{m} of about 0.42 (remember that values for repetitive contractions will be about one half of this) for a range of temperatures from 0 °C to 30 °C. Muscle efficiency then, is not temperature dependent.

Red and white muscle

The swimming activity of many fish may be divided into two broad categories; low-speed sustained cruising and high-speed burst swimming (activity levels are discussed in detail in the following chapter). A large range of power output is required of the propulsive musculature to meet these needs. Fish have evolved a 'two-geared' system. Each muscle block (myotome) consists of a relatively small quantity of fibres which are designed for cruising (red muscle) and a much larger mass (up to 90% of the total muscle mass) of fibres which are designed for burst swimming (white muscle). The locomotor muscles of birds and terrestrial vertebrates are not characterized

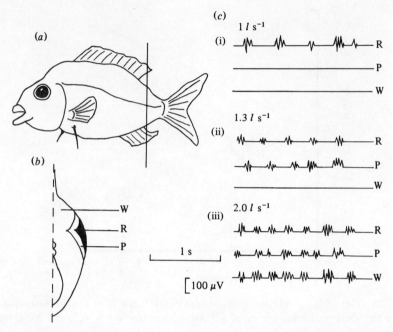

Fig. 25   Diagram of a carp (*a*) showing (*b*) the distribution of red (R), pink (P) and white (W) muscle fibres and (*c*) EMG records from the three layers at different swimming speeds.

by large distinct white muscle masses. Most fish are neutrally buoyant or nearly so, and the effective weight of the white muscle that must be supported and transported during cruising is relatively small.

In most fish the red fibres form a narrow band of muscle along the flanks of the animal (see Fig. 25*b*). Some scombriod fish, however, show a deep band of red muscle situated close to the vertebral column in addition to the superficial lateral band (see Fig. 26*b*). Red muscle fibres are characterized by a high myoglobin content, extensive capillary beds, a high content of mitochondria and oxidative enzymes (Bone, 1966, 1974, 1978). Red fibres function aerobically, fatigue slowly and have a low intrinsic velocity of shortening (Johnston, Frearson & Goldspink, 1972). These properties are required for the efficient generation of the slow repetitive movements associated with low-speed swimming.

White muscle is designed for maximum power output. It is characterized by a high myofibrillar density, a low content of mitochondria and a poor blood supply. The white fibres function anaerobically; glycogen (which is stored within the fibres) is converted to lactic acid. White fibres fatigue rapidly and have a high intrinsic speed of shortening (Flitney & Johnston, 1979).

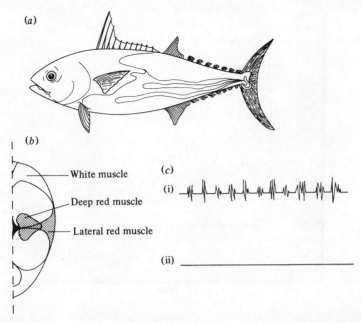

Fig. 26 (*a*) Lateral view of the skipjack tuna *Katsuwonus pelamis*. (*b*) Cross-section anterior to the dorsal fin showing the distribution of muscle fibre types. (*c*) EMG from the lateral red muscle (i) and the white muscle (ii) at low swimming speed. Modified from Rayner & Keenan (1967).

Fibres that are intermediate in biochemical and ultrastructural characteristics between red and white fibres have been identified in some fish; they are often referred to as pink fibres. Biochemical studies (Johnston, Davison & Goldspink, 1977) indicate that pink fibres have an intrinsic speed of shortening that is between that of the red and the white fibres.

Electromyographic work has demonstrated that in some fish only red muscle is active during low-speed swimming. Bone (1966) observed a clear division of activity between the red and white muscles with respect to swimming speed in the dogfish (*Scyliorhinus*). Similar results were found for the skipjack tuna (*Katsuwonus pelamis*) by Rayner & Keenan (1967). However, a clear division of labour does not exist between the two muscle types in all cases.

On the basis of a respirometric study, Smit, Amelink-Koustaal & Vijverberg (1971) concluded that there is a substantial anaerobic contribution to swimming metabolism at low speeds in the goldfish (*Carassius auratus*). Johnston & Goldspink (1973) came to a similar conclusion from lactate measurements. The carp (*Cyprinus carpio*) has a pink fibre population that is recruited after the red fibres. Finally the white fibres become active

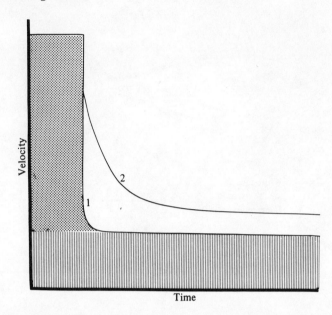

Fig. 27   Schematic endurance (velocity–time) curves. The 'barred' area represents the velocity range of the red muscle system alone, the stippled area that of the white muscle system where it is not employed in sustained swimming (curve 1). Curve 2 is to be expected when the white muscle is employed during sustained swimming. From Bone (1974).

(Johnston *et al.* 1977; see Fig. 25*c*). Bone (1974) has suggested (on the basis of electromyographic work) that white fibre activity may be local at low swimming speeds in the carp.

The form of the velocity versus endurance curve should be influenced by the extent of white and/or pink muscle involvement in low-speed swimming. A schematic representation of two possibilities is given in Fig. 27. Bone (1974) points out that the curve for jack mackerel (*Trachurus symmetricus*) obtained by Hunter (1971) is of the form to be expected when white fibres are not active at low speeds.

Relative proportions of red and white muscle fibre types can be correlated with mode of life in fish (Boddeke, Slijper & van der Stelt, 1959; Bone, 1966; Greer-Walker & Pull, 1975). Greer-Walker & Pull (1975) histologically sampled cross-sections (made at a point one-third of the way back from the snout) of a large variety of fish. Active pelagic forms (e.g. Scombriodae, Clupeidae, Carangidae) typically showed a high proportion of red muscle (about 20% of the total muscle mass). More sluggish forms (e.g. Anguillidae, Congridae, Macrouridae) had a lower proportion of red muscle.

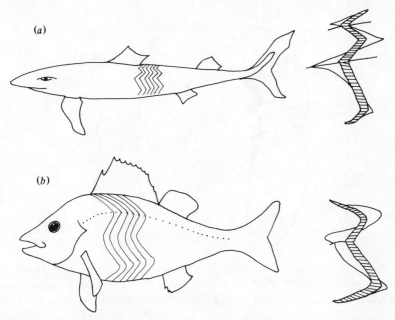

Fig. 28 Typical myomere structure in an elasmobranch (*a*) and a teleost (*b*). The diagrams on the right show enlarged views of typical trunk myotomes.

Anatomy and fibre geometry

The propulsive axial-body musculature of fish is divided into successive segments (myomeres) which are separated by sheets of connective tissue (myosepta). The myotomes are folded into complicated zig-zag patterns (see Fig. 28). Myotome structure is described in detail by Nishi (1938), Nursall (1956), Bone (1966) and Alexander (1969).

In selachians (e.g. *Scyliorhinus*) each myoseptum is folded into a series of roughly conical sections which lie one above the other and point alternately anterior and posterior. In *Scyliorhinus* there are three anterior cones and two posterior cones. The central anterior cone is the largest and is termed the main anterior cone; smaller cones are called subsidiary cones. Lateral to the apex of each cone a tendon projects from the myoseptum. The selachian arrangement of cones is illustrated in Fig. 29*a*.

In *Protopterus* the main anterior cone is bifid and the subsidiary cones are small. The primitive teleosts (e.g. *Amia, Anguilla, Salmo*) lack subsidiary cones (see Fig. 29*b*). More advanced teleosts also lack subsidiary anterior cones. The main anterior cone may or may not be bifid. The cones of advanced teleosts are not connected to tendons.

In the superficial red muscle of the myotomes of selachians and teleosts,

(a)   (b)

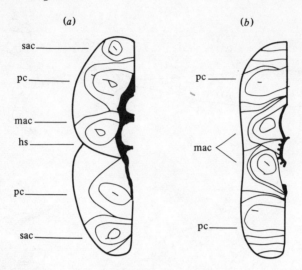

Fig. 29 Transverse sections showing the myosepta in the anterior caudal region of an advanced embryo of *Scyliorhinus* (a) and *Salmo* (b). hs, horizontal septum; mac, main anterior cone; pc, posterior cone; sac, subsidiary anterior cone. From Alexander (1969).

fibres adhere closely to the underside of the skin and connect firmly to each side of the myosepta. The red muscle is divided into segments by the myosepta. This arrangement allows the fish to produce a slow body wave as well as a variety of other slow movements (Wardle & Videler, 1980).

Some authors have assumed that the deeper white fibres are arranged in the same way as the red fibres (e.g. Jarman, 1961; Willemse, 1966). This is not so however. The white fibres are arranged in complex three-dimensional patterns with some fibres making angles of 30° or more with the long axis of the fish (Shann, 1914; Nursall, 1956; Alexander, 1969). There have been many attempts to interpret the arrangement of the deeper fibres of the myotome (e.g. Jarman, 1961; Willemse, 1966; van der Stelt, 1968; Alexander, 1969; Kashin & Smoljaninov, 1969). The account that follows is drawn from Alexander (1969).

Typical fibre trajectories approximate a segment of a helix, starting at the median plane of the fish and returning to it. The helices are arranged in coaxial bundles (Fig. 30). Fibre angles of about 35° have been measured for the outermost fibres in *Xiphophorus*. If the diameter of the fish is $2a$, the radius of the outermost trajectory is about $8a$. From these measurements Alexander calculated the pitch of the helix (Pitch $= 2\pi a \cot \theta$, where θ is the helical angle) to be about $7a$, with a length of about $4a$.

We would expect that the white fibres would contract at the rate at which maximum power is developed. If in bending the body the outermost fibres

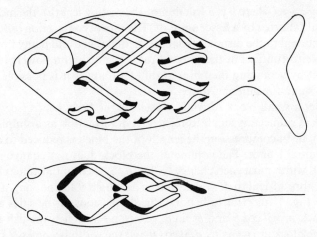

Fig. 30 Diagrammatic dorsal and lateral views of a typical teleost showing some muscle fibre trajectories. From Alexander (1969).

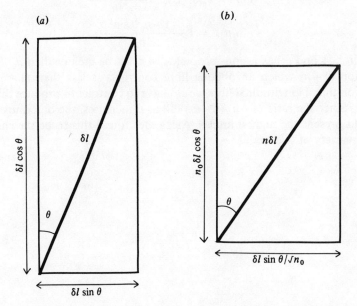

Fig. 31 White muscle deformation. This figure is explained in the text. From Alexander (1969).

shorten by $1-n$ (where n is a fraction of the resting length), the more medial fibres would shorten by a lesser amount. The rate of contraction (dn/dt) would be proportional to the distance from the medial plane and many fibres would not be able to contract at the optimal rate for maximum power. It is shown below that by arranging the fibres at different angles it is possible for dn/dt to be the same for different fibres.

Consider a small block of white muscle as the body bends (see Fig. 31a). The block is sufficiently small to regard its deformation as a simple uniaxial compression. In compression the length of the block is reduced to a fraction n_0 of its initial length. The volume of the block does not change and so its height and width must each change to $1/\sqrt{n_0}$ of their initial values. Consider a muscle fibre of length δl which makes an angle θ to the long axis of the body (see Fig. 31b). Prior to bending the dimensions of the sides and width of the block are $\delta l \cos \theta$ and $\delta l \sin \theta$, respectively. When the fish bends the sides of the block are given by $n_0 \delta l \cos \theta$ and the width becomes $\delta l \sin \theta / \sqrt{n_0}$. If the fibre has contracted to a fraction n of its initial length, then by Pythagoras' theorem

$$n^2 \delta l^2 = n_0^2 \delta l^2 \cos^2 \theta + \frac{\delta l^2 \sin^2 \theta}{n_0} \tag{54}$$

and

$$(n/n_0)^2 = 1 + \frac{(1-n_0^3)\sin^2 \theta}{n_0^3} \tag{55}$$

If $n_0 > 1$ and θ has a non-zero value, n must be greater than n_0. So, the fraction $(1-n)$ which an oblique fibre contracts is less than $(1-n_0)$, the fraction that a longitudinal fibre would have to contract to produce the same effect. Since the ratio $(1-n)/(1-n_0)$ fall as θ increases, the outermost fibres should present the highest angles. Alexander found this to be the case in a large variety of teleosts.

3

Speed, size and power

Introduction

Activity levels in fish are defined on the basis of swimming velocity–time curves. Three basic levels are recognized: sustained, prolonged, and burst swimming. Swimming speed is then discussed in relation to size (length). Data for non-scombriod and scombriod fish are summarized graphically for given activity levels. In the case of sustained swimming the curves are compared to that for optimal energetic efficiency. Burst swimming performance is discussed in relation to possible limits to swimming speed set by the contraction time of the propulsive musculature.

Other biological constraints on performance (e.g. influence of sex and disease) are not discussed. Discussion of environmental influences (e.g. effects of temperature, oxygen and carbon dioxide levels, salinity) is also omitted. Readers interested in these aspects are referred to a recent review by Beamish (1978).

Speed and stamina is discussed in freshwater and marine fish. Simple equations describe time to exhaustion as a function of swimming speed. There is a great deal of interspecific variation in the speed–time curves.

Finally, swimming speed and size is considered in relation to power requirements and Gray's Paradox. Gray (1936) investigated the swimming performance of porpoises and dolphins. He employed the standard hydrodynamic equations of resistance and assumed that a unit mass of cetacean muscle was capable of delivering the same power as that from man. Gray concluded that in order to achieve their reported swimming speeds, the animals would have to produce about seven times more power than the calculated values or be able to maintain a laminar boundary layer (Gray's Paradox). Employing modern estimates of the maximum power output of vertebrate muscles and considering some recent developments in the analysis of swimming drag, it is concluded that Gray's Paradox only applies to the burst-swimming performance of scombriods.

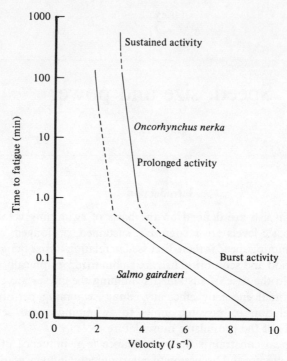

Fig. 32   Fatigue curves for *Oncorhynchus nerka* and *Salmo gairdneri*. This figure is explained in the text. (Modified after Brett, 1964, 1967.)

Activity levels

Activity levels in fish are commonly defined on the basis of the time for which a given speed can be maintained. Three basic swimming levels can be recognized: (1) sustained; (2) prolonged; and (3) burst. These levels of activity can be defined on the basis of a velocity–time (endurance) curve. Endurance curves for sockeye salmon (*Oncorhynchus nerka*) and rainbow trout (*Salmo gairdneri*) are shown in Fig. 32. In both cases the endurance curve can be divided into three straight lines corresponding to the main activity levels.

For practical purposes sustained swimming is usually taken to mean any swimming behaviour that is maintained for more than 200 min (Brett, 1967). Sustained swimming includes low levels of routine activity (e.g. territorial behaviour, foraging and station holding), schooling (including migration) and cruising. The speeds at which negatively buoyant fish (e.g. scombriods, xiphoids) must swim in order to maintain their level in the water are included in the subcategory of cruising. Metabolism is aerobic during sustained swimming.

Prolonged swimming is defined as activity that is maintained for between

200 min and 15 s. It commonly takes the form of periods of cruising interspersed with bouts of more vigorous swimming. Energy is supplied through aerobic and/or anaerobic pathways. Many laboratory based observations have been made on prolonged swimming activity in fish (e.g. Brett, 1964, 1967; Webb, 1971*a*, *b*). Experiments in water flumes usually involve a series of increasing velocity tests. Brett (1964) introduced the concept of 'critical swimming speed' (U_{crit}) to accurately quantify prolonged activity in fish. Essentially, swimming speed is increased incrementally by an amount ΔU every t min; if a fish is exhausted in t' min following a speed increase from U to $U + \Delta U$, then the t-minute U_{crit} is given by

$$U_{crit} = \left(\frac{\Delta U t'}{t}\right) + U \tag{56}$$

Burst swimming involves high-speed movements of relatively short duration (< 15 s). Activity can be steady or unsteady. Fast-starts and rapid turns (see chapter 5) are included in this category. Accelerations of the order of 40–50 m s^{-2} have been measured during the fast-start behaviour of fish (Hertel, 1966; Weihs, 1973*a*; Webb, 1975*a*, 1978, 1981). Metabolism is anaerobic during burst swimming.

Size and speed

During the past 20 years or so a great deal of work has been done on the swimming speeds of fish in relation to length. Observations on non-scombriod fish are reviewed in detail by Blaxter (1969) and Beamish (1978). Magnuson (1978) has summarized the data on scombriods.

Data on the prolonged swimming performance of a large variety of non-scombriod fish are plotted in Fig. 33. Fig. 34 gives the data on scombriod fish. The slopes of the regression lines for non-scombriod and scombriod fish are similar (0.81 and 0.89, respectively); however, the variability about the lines is large. Also plotted in Figs 33 and 34 are the curves corresponding to the optimal swimming speed for minimum energy expenditure (Weihs, 1978). The curve falls well below the line of best fit for the experimental data in both cases. The slope of the curve (0.5) is significantly different from that of the experimental curves.

Data on the burst swimming activity of non-scombriods and scombriods are plotted in Figs 35 and 36, respectively. For many years it was thought that 10 body lengths per second (10 l s^{-1}) represented a reasonable upper limit to the burst swimming performance of fish irrespective of their size. Although the regression lines in Figs 35 and 36 fall close to the line corresponding to the 10 l s^{-1} 'rule of thumb', the variation about the regression lines is large and many of the observations exceed 10 l s^{-1}. Recently values of up to 25 l s^{-1} have been reported for small (< 0.1 m) fish (Wardle, 1975).

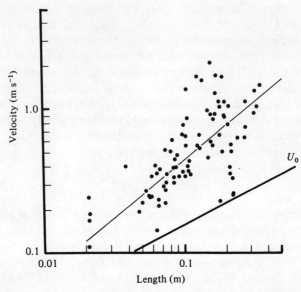

Fig. 33   Logarithmic plot of swimming velocity against length for prolonged swimming activity in non-scombriod fish. Data from Blaxter (1969) and Beamish (1978).

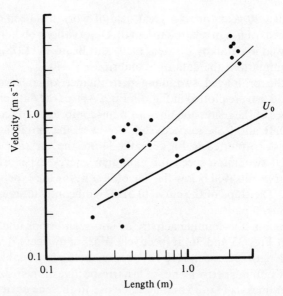

Fig. 34   Logarithmic plot of swimming velocity against length for sustained swimming in scombriod fish. Data from Magnuson (1978).

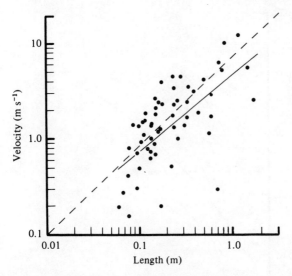

Fig. 35　Logarithmic plot of burst-swimming speed against length for non-scombriod fish. The line of identity (−−−) corresponds to $10\,l\,\mathrm{s}^{-1}$. Data from Beamish (1978).

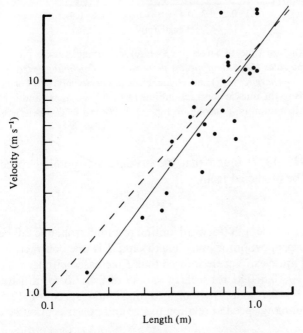

Fig. 36　Logarithmic plot of burst-swimming speed against length for scombriod fish. The solid line is a regression line for the points. The line of identity (−−−) corresponds to $10\,l\,\mathrm{s}^{-1}$. Data from Magnuson (1978).

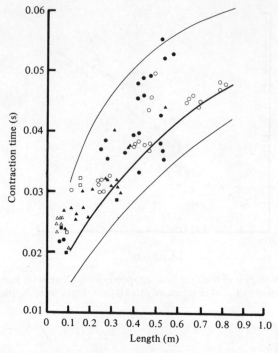

Fig. 37   Contraction time of anaerobic (white) swimming muscle plotted against length for *Salmo salar* (●), *Pleuronectes platessa* (▲), *Trachinus vipera* (□), *Gadus morhua* (○), *Clupea harengus* (△) and *Melanogrammus aeglefinus* (■). The heavy dark line corresponds to the muscle contraction time at 14 °C, the upper and lower curves correspond to the minimum contraction time at 20 °C and 0 °C, respectively. From Wardle (1975).

Wardle (1975, 1977) found that the maximum swimming speed of fish (U_{max}) could be predicted from

$$U_{max} = \frac{Yl}{2T} \tag{57}$$

where Y is the stride length (forward motion resulting from one tail beat cycle expressed as a proportion of body length) and T is one contraction time of the swimming muscles. Wardle showed that T increases with length (see Fig. 37) and that $Y = 0.6$–0.81 for a large variety of fish. So for a fish 0.1 m in length with $Y = 0.7$ and $T = 0.02$ s, a maximum speed of $17.5\,l\,s^{-1}$ is predicted. Swimming speed in relation to size and power is discussed later in this chapter.

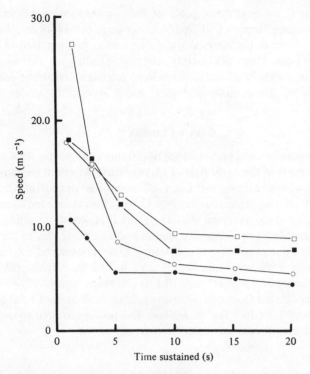

Fig. 38 Maximum sustained speeds for different periods of time by rainbow trout (*Salmo gairdneri*) of 0.28 m (□), 0.20 m (■), 0.15 m (○) and 0.10 m (●) in length. Redrawn from Bainbridge (1960).

Speed and stamina

Bainbridge (1960) studied speed and stamina in dace (*Leuciscus leuciscus*), trout (*Salmo irideus*) and goldfish (*Carassius auratus*). Maintained speed in these forms falls from about 10 l s^{-1} for activity lasting for about 1 s to about 4 l s^{-1} for activity lasting for about 20 s. Examples of velocity–duration curves for trout are given in Fig. 38. When the data is plotted on logarithmic co-ordinates straight lines are obtained and

$$\log (\text{time to exhaustion}) = a + k \log (\text{velocity}) \tag{58}$$

The ability to sustain periods of swimming is proportional to $l^{-1.09}$ and $l^{-0.65}$ (i.e. $k = -1.09$ and -0.65) for dace and trout, and goldfish, respectively. Bainbridge suggests that the difference between the two groups could be due to a combination of differences between the two groups involving muscle mass, fineness ratio and other physiological factors.

Dorn, Johnson & Darby (1979) have investigated the swimming ability of nine species of Californian inshore fishes in a flume over a range of velocity of $0.42–1.17$ m s^{-1}. Logarithmic plots of time to exhaustion–velocity gave values of k ranging from -1.91 in *Hypsurus caryi* to -4.48 in *Phanerdon furcatus*. The extent of the interspecific differences is striking. Before we will be able to explain them adequately, detailed studies on the swimming performance of a variety of fish in relation to muscle activity, the geometric characteristics of the animals and their exact swimming modes will be required.

Gray's Paradox

The exceptional swimming performance of dolphins and porpoises was noted by many observers in the early part of this century. At about the same time physiologists were obtaining estimates of the power output of vertebrate striated muscles during strenuous activity. Gray (1936) attempted to calculate the power required to overcome the drag force that would be acting on the animals at their observed swimming speeds and to equate this to the power available from their propulsive musculature. Gray considered the case of a 1.22 m porpoise (*Phocaena communis*) and a 1.82 m dolphin (*Delphinus delphus*) swimming at 7.6 m s^{-1} and 10.1 m s^{-1}, respectively.

Drag was calculated from eqn 34, using a drag coefficient of 1.5×10^{-3} for *P. communis* and 1.3×10^{-3} for *D. delphus*. The power required to overcome the calculated drag forces was calculated from

$$P_{\text{req}} = \tfrac{1}{2}\rho S_w U^3 C_f \tag{59}$$

Gray calculated values of 448 W and 1940 W for *P. communis* and *D. delphus*, respectively. These values were equated with estimates of the power available from the propulsive musculature. The specimen of *P. communis* weighed 24 kg of which 4.05 kg was propulsive muscle. The specimen of *D. delphus* was much larger, with a mass of 91 kg of which 15.9 kg was estimated to be propulsive muscle.

A figure for the specific power output of the muscle was taken from estimates of the power output of a highly trained oarsman. Henderson & Haggard (1925) estimated that each oarsman in a rowing eight would produce about 373 W. About 22.5 kg of muscle was employed, giving a specific power output of approximately 16.6 W kg^{-1} of muscle. Assuming that the specific power output of a man and a cetacean are about the same, the 1.22 m porpoise would have required 27 kg of propulsive musculature to swim at the observed speed. The dolphin would have required 117 kg of muscle. These values are about seven times those of the actual muscle masses. This then is Gray's Paradox.

The values for the drag coefficients were based on a turbulent boundary layer. If the boundary layer were assumed to be laminar the values of the drag coefficients would be smaller (about 2.3×10^{-4} for the porpoise and about

1.5 × 10⁻⁴ for the dolphin) and the power requirements correspondingly less. Gray calculated that the porpoise would need about 67.1 W, and that the dolphin would need approximately 223 W to swim at the observed speeds, if the boundary layer flow were to be laminar in both cases. These figures are consistent with the actual amounts of propulsive musculature present in the animals. However, this implies that cetaceans are able to maintain a laminar boundary layer at very high Reynolds Numbers (10^6–10^7). Alternately, the paradox could be resolved by assuming that the muscles of cetaceans are capable of delivering about seven times as much power per unit mass as human muscles.

On the basis of the results of a series of bicycle ergometer studies by Dickinson (1928), Parry (1949) suggested that a value of 40 W kg⁻¹ for the specific power output of vertebrate muscle would be appropriate. He noted that this increase in power would not be sufficient to resolve the paradox.

Gero (1952) recorded a velocity of 12.2 m s⁻¹ for a 1.22 m barracuda (*Sphyraena barracuda*). Employing similar methods to Gray, but assuming that 50–75% of the surface of the fish experienced laminar flow, Gero calculated a maximum required power output of 604 W. This corresponds to a specific power value in the region of 240–330 W kg⁻¹. To many physiologists at the time these figures seemed too high. Efforts to resolve the paradox began to centre on discovering a mechanism by which cetaceans and certain fish could maintain a laminar boundary layer at high Reynolds Numbers.

Kramer (1960*a, b*) experimentally investigated the hydromechanical properties of an analogue of dolphin skin. The analogue consisted of a rubber diaphragm supported by studs. The spaces between the studs were filled with a viscous fluid. At low Reynolds Numbers the system behaved like a rigid sheet, but at high Reynolds Numbers disturbances in the water were transmitted to the viscous cells and the drag on the system was reduced. However, the reduction was not sufficient to indicate the presence of a laminar boundary layer over the entire system.

Bainbridge (1961) makes some important modifications to Gray's analysis. Firstly, he includes a pressure drag term based upon fineness ratio (see eqn 39). For a large variety of fish Bainbridge shows that the pressure drag term will be about 20% of the skin friction drag, and so the power required becomes

$$P_{\text{req}} = \tfrac{1}{2}\rho S_{\text{w}} U^3 1.2 C_{\text{f}} \tag{60}$$

Secondly, he points out that at any instant only one-half of the propulsive musculature will be active. The power available can be written as

$$P_{\text{avail}} = \frac{P_{\text{f}} M_{\text{w}}}{2} \tag{61}$$

where P_{f} is a power factor and M_{w} is the mass of the propulsive musculature.

Thirdly, a propeller efficiency term of 75% is introduced. Bainbridge equates the power required to overcome swimming drag to the power available

$$\tfrac{1}{2}\rho S_{\rm w} U^3 1.2 C_{\rm f} = 0.75 P_{\rm f} \frac{M_{\rm w}}{2} \tag{62}$$

Assuming that $S_{\rm w} = 0.41^2$ (Bainbridge, 1960) we can write

$$0.64 l^2 U^3 C_{\rm f} = P_{\rm f} M_{\rm w} \tag{63}$$

The expressions for the frictional drag coefficient for laminar and turbulent boundary layer flow (see eqns 36 and 37) can be written in the following form

$$C_{\rm f(lam)} = \frac{1.33 \nu^{\frac{1}{2}}}{l^{\frac{1}{2}} U^{\frac{1}{2}}} \tag{64}$$

$$C_{\rm f(turb)} = \frac{0.074 \nu^{\frac{1}{5}}}{l^{\frac{1}{5}} U^{\frac{1}{5}}} \tag{65}$$

Substitution of eqns 64 and 65 into eqn 63, followed by rearrangement gives

$$U_{\rm lam} = \left(\frac{P_{\rm f} M_{\rm w}}{0.085\, l^{\frac{3}{2}}}\right)^{\frac{2}{5}} \tag{66}$$

$$U_{\rm turb} = \left(\frac{P_{\rm f} M_{\rm w}}{0.01885\, l^{\frac{9}{5}}}\right)^{\frac{5}{14}} \tag{67}$$

(Bainbridge, 1961). Assuming that about 50% of the mass of the fish is propulsive musculature, Bainbridge derived the expression $M_{\rm w} = 0.005 l^{2.9}$. Substituting this and a power factor of 40 W kg^{-1} into eqns 66 and 67 gives

$$U_{\rm lam} = 56.06 l^{0.56} \tag{68}$$

$$U_{\rm turb} = 62.35 l^{0.39} \tag{69}$$

Eqns 68 and 69 can be employed to predict the swimming speed of fish of a given length for the cases of laminar and turbulent boundary layer flow, respectively. Bainbridge's curves are reproduced in Fig. 39, where they are compared with experimental observations on the sustained swimming speed of a variety of small (up to 0.4 m in length) fish. All but one of the experimental points falls below the iso-Reynolds Number line corresponding to transition on a streamlined body, and therefore laminar boundary layer flow is expected. The points are well below the predicted performance curves, and so Gray's Paradox does not apply to small fish during sustained swimming.

In Fig. 40 experimental observations on burst-swimming speeds in small fish are compared with the predicted performance curve for the case of a turbulent boundary layer. Many of the points fall above the curve and the iso-Reynolds Number line for transition. Gray's Paradox may thus apply in this case. When the Bainbridge estimates are applied to larger fish (scombriods, etc.) and cetaceans (porpoises and dolphins) Gray's Paradox also applies.

During the past 20 years great advances have been made in muscle

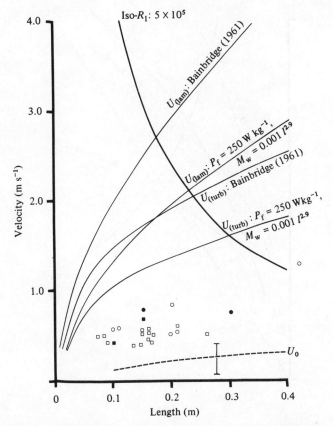

Fig. 39 Predicted performance (velocity–length) curves for sustained swimming in small fish. Symbols refer to experimental observations by Bainbridge (1960) on *Leuciscus leuciscus* (□), *Salmo irideus* (○), *Carassius auratus* (■) and *Salmo gairdneri* (●). The bar refers to a range of speeds sustained by *S. gairdneri* (Webb, 1971a). The curve corresponding to optimal energetic advantage (U_0) is also shown (Weihs, 1977).

physiology and in our understanding of the hydrodynamics of fish swimming. A new series of estimates and assumptions concerning the swimming performance of fish is now appropriate. We will assume: (1) that during sustained swimming only red muscle is active and that 10–20% of the total muscle mass is red muscle; (2) that in burst swimming only, white muscle is active and that 80–90% of the total muscle mass is white muscle; (3) maximum specific power outputs of 250 W kg^{-1} and 500 W kg^{-1} for red and white muscle, respectively (Weis-Fogh & Alexander, 1977); and (4) that the skin friction drag of an oscillating fish is about four times that of the equivalent rigid body (Lighthill, 1971). In scombriods, oscillation of the body

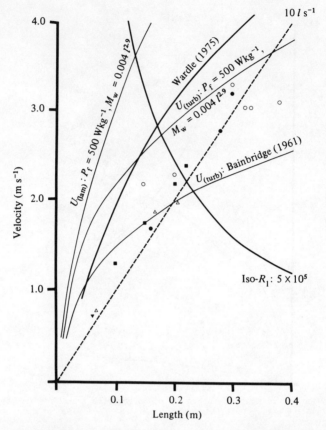

Fig. 40   Calculated performance curves for burst swimming in small fish. The points are from Bainbridge (1960) and refer to *Leuciscus leuciscus* (■), *Salmo gairdneri* (●), *Carassius auratus* (△) and *Salmo irideus* (○). The curve corresponding to the limit to swimming speed based on muscle contraction time (Wardle, 1975) is also shown.

is restricted to a rigid caudal fin, and therefore the 'equivalent rigid drag' is employed. The rigid-body analogy in fish swimming is discussed in detail in chapters 4 and 5.

So,

$$\tfrac{1}{2}\rho S_\mathrm{w} U^3 4.2 C_\mathrm{f} = 0.75 P_\mathrm{f} \frac{M_\mathrm{w}}{2} \tag{70}$$

Substituting eqns 68 and 69 into eqn 70 and rearranging gives

$$U_\mathrm{lam} = \left(\frac{P_\mathrm{f} M_\mathrm{w}}{0.297\, l^{\frac{3}{2}}}\right)^{\frac{2}{5}} \tag{71}$$

$$U_\mathrm{turb} = \left(\frac{P_\mathrm{f} M_\mathrm{w}}{0.0657\, l^{\frac{8}{5}}}\right)^{\frac{5}{14}} \tag{72}$$

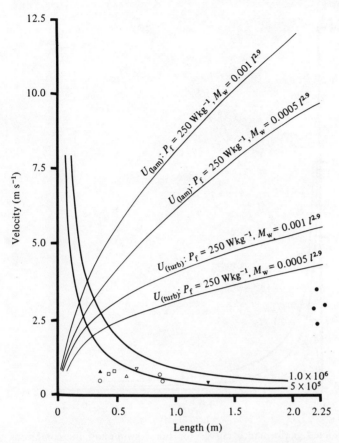

Fig. 41 Calculated performance curves for sustained swimming speed in scombriods and experimental points for *Thunnus thynnus* (●), *T. alalunga* (○), *T. obesus* (△), *T. ablacores*, (▲), *Acanthocybium solander* (▼) and *Euthynnus pelamis* (□). Data from Magnuson (1978). Iso-Reynolds Number curves for 5×10^5 and 1×10^6 are also shown.

For $M_w = 0.001 l^{2.9}$ (i.e. 20% red muscle) and $P_f = 250$ W kg^{-1}, we have

$$U_{\text{lam}} = 37.16 l^{0.56} \tag{73}$$

$$U_{\text{turb}} = 43.1 l^{0.39} \tag{74}$$

and for $M_w = 0.004 l^{2.9}$ (i.e. 80% white muscle) and $P_f = 500$ W kg^{-1}

$$U_{\text{lam}} = 85.37 l^{0.56} \tag{75}$$

$$U_{\text{turb}} = 90.7 l^{0.39} \tag{76}$$

For scombriod fish, eqns 66 and 67 apply with the new values of M_w and

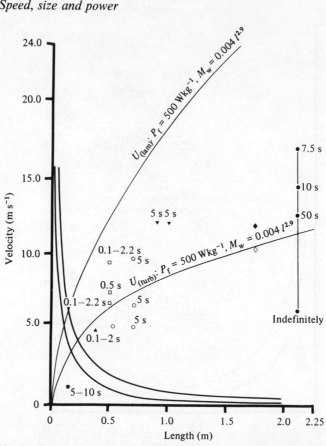

Fig. 42   Predicted performance curves for burst swimming in scombrioids and experimental points for *Sarda sarda* (■), *Euthynnus affinis* (▲), *E. pelamis* (□), *Thunnus albacores* (○), *Acanthocybium solanderi* (▼), a dolphin (◇), a barracuda (◆) and a series of speeds for *Tursiops gilli* (●). From Gray (1936), Gero (1952), Lang (1974) and Magnuson (1978). Iso-Reynolds Number Curves for 5×10^5 and 1×10^6 are also shown. The duration of swimming in *Tursiops* is indicated.

P_f. For M_w equal to 10% of the total muscle mass and $P_f = 250$ W kg^{-1}

$$U_{lam} = 46.45 l^{0.56} \tag{77}$$

$$U_{turb} = 52.64 l^{0.39} \tag{78}$$

For the case appropriate to burst swimming (80% white muscle active with a power factor of 500 W kg^{-1})

$$U_{lam} = 140.8 l^{0.56} \tag{79}$$

$$U_{turb} = 141.6 l^{0.39} \tag{80}$$

It can be shown that

$$U_{lam} = 1.625 P_f^{0.4} M_w^{0.4} l^{0.56} \tag{81}$$

$$U_{turb} = 2.64 P_f^{0.357} M_w^{0.357} l^{0.39} \tag{82}$$

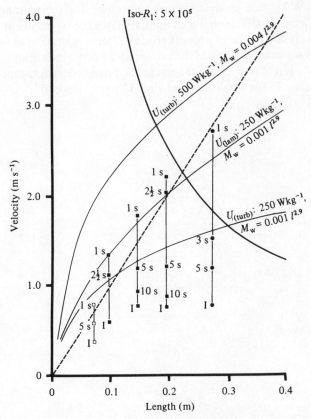

Fig. 43 Predicted performance curves and observations on the swimming speeds of *Leuciscus leuciscus* (■), *Carassius auratus* (□) and *Salmo irideus* (●) maintained for various periods of time. The experimental data is from Bainbridge (1960). The Iso-Reynolds Number Curve for 5×10^5 is indicated.

Eqns 73 and 74 are plotted in Fig. 39. Although the new curves fall below those of Bainbridge (1961), they are both well above the experimental points for sustained swimming in small fish. The curves for burst swimming in these fish (eqns 75 and 76) lie above the corresponding curves of Bainbridge (see Fig. 40), and therefore Gray's Paradox no longer applies to small fish in burst swimming.

For larger fish (scombriods) in sustained swimming, the experimental observations fall far below the new curves and Gray's Paradox does not apply (Fig. 41). However, in order to account for the burst swimming performance of the scombriods it is necessary to postulate laminar boundary layer flow (Fig. 42).

Kermack (1948) has shown that Gray's low estimate of specific power output (16.6 W kg^{-1}) is easily sufficient to account for the swimming performance of large whales, even if the boundary layer flow is turbulent. Whales have a relatively large mass of propulsive musculature relative to their surface (muscle mass scales as l^3, whereas the surface area upon which drag acts scales as l^2). The relation of speed and stamina to the predicted performance curves is shown in Fig. 43 for small fish and Fig. 42 for a cetacean.

4

Drag and drag reduction mechanisms

Introduction

This chapter begins by considering some of the experimental methods that are available for determining the dead drag of fish. The physical basis of dead-drag measurements is outlined and the apparatus employed in terminal velocity, deceleration in glide, towing-tank and wind and water tunnel experiments is described. Values of the drag coefficient obtained from the various experimental methods for streamlined and unstreamlined fish are discussed in relation to theoretical values calculated from the standard hydrodynamic equations of resistance. Generally the experimental values far exceed theoretical values. Possible explanations for the discrepancies are offered and discussed.

Evaluation of the results of dead-drag experiments leads us on to consider the validity of the rigid-body analogy, which assumes that the drag of a dead fish is at least a good approximation to that of a swimming fish. It is concluded that the rigid-body analogy is not likely to apply to fish that swim by undulating most of their body because of additional pressure and friction drag terms associated with repeated bodily flexure. Certain fish are propelled by the action of discrete median and paired fins. The body is rigid, and therefore dead-drag measurements probably give a good indication of true swimming drag in these cases.

The second part of this chapter is concerned with the function of possible drag reduction mechanisms in fish. The initial stimulus for research in this area was provided by Gray's Paradox. We now know that the swimming performance of most fish can be explained without the need to postulate any special drag reducing mechanisms (see chapter 3), yet over the past 30 years or so various examples have been identified in the faster swimming fish (e.g. the scombriods). Viewing drag reduction mechanisms as adaptations that reduce the energy cost of transport and increase the limits of swimming speed, we can expect them to be selected for in evolution.

Mechanisms directed at maintaining a laminar boundary layer are discussed first. Overall body form, the possible damping properties of cetacean skin, some possible boundary layer suction devices and the role of gill effluent in boundary layer control are discussed. Although there is no experimental evidence it seems likely that these mechanisms act to reduce drag, at least in some cases. The possibility that fish might be able to actively sense boundary layer disturbances and damp them out is considered to be highly unlikely.

The possible role of mucus in reducing frictional drag through reducing the viscosity of the boundary layer is discussed. Experimental work on mucus–water mixtures indicates that frictional drag may be reduced by as much as 60% in some cases. The use of metabolically generated heat to directly reduce the viscosity of the boundary layer is considered to be unlikely.

Methods of inducing turbulent boundary layer flow in order to prevent or delay transition and separation are considered next. The role of roughness elements such as spines and scales and the effect of direct injections of momentum into the boundary layer are discussed. Finally, the possible influence of structures such as finlets and keels on boundary layer flow and flow outside the boundary layer is discussed.

Possible behavioural mechanisms that might influence the swimming drag of individual fish (e.g. schooling, burst–glide swimming) are considered in chapter 8.

Measurement of drag forces

Terminal velocity experiments

Free-fall measurements have been employed to measure the dead drag of a variety of fish (Magnan & St Laque, 1929; Magnan, 1930; Richardson, 1936; Gero, 1952; Blake, 1979a, b, 1981a, c). In typical experiments dead fish are dropped from the jaws of an electronically controlled holding device to terminal velocity down a large vertical tank of water (see Fig. 44). Two basic methods can be employed to determine the form of the distance–time curve: direct chronographic recording (e.g. Richardson, 1936); and filming the fish against a background grid as it falls (e.g. Gero, 1952; Blake, 1979a, b, 1981a, c).

The equation of motion of a falling fish may be written as

$$m\frac{dU}{dt} = mg - kU^n \tag{83}$$

where m is the mass of the fish and any water that it entrains due to its unsteady motion (typically about 10–20% of the actual mass of the fish) and kU^n is the drag force. In practice, n may have values between 0 and 2. For $n = 1$

$$U = \frac{mg}{k}(1 - e^{kt/m}) \tag{84}$$

(Stoke's Law) and the terminal velocity (U_t) is

$$U_t = \frac{mg}{mk} \tag{85}$$

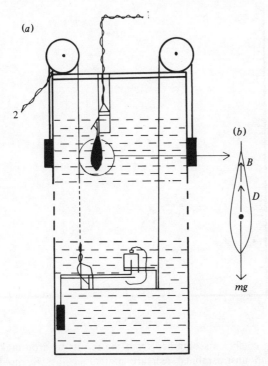

Fig. 44 (*a*) Drop-tank apparatus. Cables connected to the starting switch (1) and stopping switch (2) of a chronograph are indicated. (*b*) Fish-shaped body showing the forces acting during descent in the drop-tank (**B**, buoyancy; D, drag; mg, weight). Partly based on Richardson (1936).

For $n = 2$
$$U = \frac{(mg/mk)^{\frac{1}{2}} \tanh g}{mg/mk} \qquad (86)$$

(Newton's Law) and
$$U_{\mathrm{t}} = \left(\frac{mg}{mk}\right)^{\frac{1}{2}} \qquad (87)$$

Magnan & St Laque (1929) performed drop-tank experiments on a wide variety of fish. However, they do not provide sufficient information for the calculation of drag coefficients. Magnan (1930) gives a series of distance–time curves for the free-fall of a variety of fish and concludes that the curves are exact parabolas (i.e. $n = 0$). This implies that acceleration was constant and that drag is independent of velocity, circumstances that only arise for free-fall in a vacuum.

Values of the drag coefficients obtained by Richardson (1936) and by Gero (1952) are about two orders of magnitude higher than values calculated from the standard hydrodynamic equations of resistance (see Fig. 45). Webb

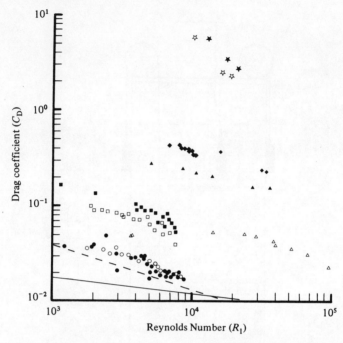

Fig. 45   Drag coefficients obtained from the results of drop-tank experiments on streamlined and unstreamlined fish are plotted against Reynolds Number. The symbols refer to angelfish without their pectoral fins and with the paired fins in the 'open' position ((●) and (■), respectively), blue gourami ((○), pectoral fins absent; (□), pectoral fins open), electric fish (△), seahorse (▲), boxfish (◆), mackerel (★) and herring (☆). The theoretical curves for the case of laminar (broken line) and turbulent (solid line) boundary layer flow are indicated. Data from Richardson (1936), Gero (1952), Blake (1979a, b, 1981a, c).

(1975a) suggests that the fish involved in the Richardson and Gero experiments were not allowed to reach terminal velocity, leading to an overestimation of the drag coefficients. It is also possible that body and fin flutter increased the drag force acting on the fish, thereby decreasing the terminal velocity. Body oscillation can be substantially reduced by stiffening the fish with steel wire.

Blake (1979a, b, 1981a, c) gives the results of a series of drop-tank studies on angelfish (*Pterophyllum eimekei*) and blue gourami (*Trichogaster tricopterus*). The results are plotted in Fig. 45. Both species are well streamlined and the experimental points are close to the theoretical curve based on a laminar boundary layer at the same Reynolds Numbers. Results for some less well streamlined fish are also given in Fig. 45.

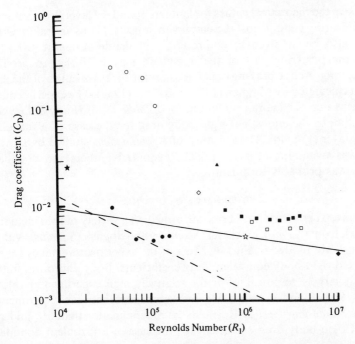

Fig. 46   Drag coefficient versus Reynolds Number for dead-drag measurements on streamlined fish. The symbols refer to *Scomber* (●), *Ambramis* (★), *Amia* (☆), *Esox* (■), *Mustelus* (▲), Salmon, (◇), *Salmo* (□) and *Lagenorhynchus* (◆). The theoretical curves for laminar (broken line) and turbulent (solid line) boundary layer flow are indicated. Based on Webb (1975*a*).

Deceleration in glide

This technique has been employed by Magnan (1930) and Gray (1957, 1968) on small fish and Lang & Daybell (1963) for a porpoise. In typical experiments fish are filmed over a grid as they passively glide after a period of active swimming. Distance–time curves are constructed and velocity calculated as a function of time. We can write

$$m\frac{dU}{dt}(t) = -\tfrac{1}{2}\rho U^2 S_w C_D(t) \tag{88}$$

from which C_D can be calculated. During the glide a mass of entrained water must be decelerated, and so m contains an added mass term.

Values of the drag coefficient calculated for bream (*Ambramis*) (Magnan, 1930), salmon (*Salmo*) (Gray, 1957) and porpoise (*Lagenorhynchus obliquidens*) (Lang & Daybell, 1963) are plotted in Fig. 46. The points for *Ambramis* and *Salmo* correspond to subcritical Reynolds Numbers and therefore boundary layer flow should be laminar. However, the points for these fish lie well above the theoretical values for laminar boundary layer flow predicted by eqn 36. During gliding most fish employ their pectoral fins as brakes. Other fins may also be involved in controlling and decelerating the fish. For example, the posterior part of the dorsal fin may be furled to produce additional drag. Active braking could account for the high values of the drag coefficient calculated by Magnan (1930) and Gray (1957). The experimentally determined value of the drag coefficient for *L. obliquidens* (Lang & Daybell, 1963) is close to the equivalent rigid-body drag for the case of a turbulent boundary layer (Fig. 46). The specimen of *L. obliquidens* studied by Lang & Daybell was swimming at supercritical Reynolds Numbers and boundary layer flow was probably turbulent.

Towing-tank experiments

Towing-tank experiments have been performed on a variety of fish, including pike (*Esox*), trout (*Salmo*), mackerel (*Scomber*) and dogfish (*Squalus*). Values for these fish are plotted in Fig. 46. Most of the experimental values exceed the predicted rigid-body equivalent drag coefficients. Body and/or fin flutter is probably largely responsible for the relatively high experimental values. Sundnes (1963) found a reasonable correspondence between experimental values for *Salmo gairdneri* at Reynolds Numbers greater than 10^6 and the equivalent rigid-body drag coefficient for the case of a turbulent boundary layer.

Water and wind tunnel measurements

Water and wind tunnels have not been used extensively for making dead-drag measurements on fish. Wind tunnels have the obvious disadvantage that the fish is in air and must be frozen to prevent fluttering and to facilitate secure attachment to a force balance. Harris (1936) measured the drag acting on a model of the smooth dogfish (*Mustelus canis*) in a wind tunnel at subcritical Reynolds Numbers. The experimental values far exceeded the equivalent rigid-body values for the case of a laminar boundary layer (Fig. 46).

Blake (1980*b*) employed a low-speed open-circuit wind tunnel to measure the drag force on a frozen specimen of the electric fish (*Gymnarchus niloticus*) and a dried seahorse (*Hippocampus hudsonius*). Values for *G. niloticus* were well above the theoretical values for a laminar and a turbulent boundary layer (Fig. 45). Experimental values for *H. hudsonius* were also high, however *Hippocampus* is a poorly streamlined fish and high values of the drag coefficient are to be expected.

Webb (1975*a*) has summarized the results of a series of water tunnel studies (Brett, 1963; Webb, 1970; Mearns in Webb, 1975*a*) on the dead drag of Salmonids. The effects of body rigidity, fin flutter and size on drag are summarized in Fig. 47. High rigidity, the absence of the paired fins and large size are associated with minimum dead drag. High drag coefficients are associated with low rigidity, fins that are free to flutter and small size.

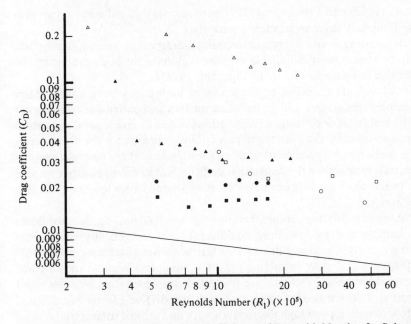

Fig. 47   Relation between dead-drag coefficient and Reynolds Number for Salmonids. Values were determined from the results of force measurements made in a flume. Open triangles refer to a small specimen of *Oncorhynchus nerka* which was partially stiffened with wire. The solid triangles are for medium-sized rainbow trout which were stiffened with wire and held rigid along their centre-line. Solid circles correspond to medium-sized trout of high rigidity. These fish were supported with additional tension bars that acted to further reduce flutter. The solid squares refer to medium-sized trout stiffened with wire and tension bars with no pectoral fins. Open circles correspond to large coho salmon and trout of low rigidity. The solid line shows the theoretical values of the drag coefficients for turbulent boundary layer flow. From Webb (1975*a*).

The rigid-body analogy

Experimental determinations of the dead drag of a rigid fish or fish model are only likely to give a good estimate of swimming drag if it can be assumed that the flow pattern around a fish during a dead-drag determination is similar to that produced by a swimming fish. For many common pelagic, fusiform fish that are propelled by repeated oscillation of their body and caudal fin (undulatory propulsion is discussed in the following chapter) this is unlikely, as:

1. Experimental evidence indicates that the pressure drag of a fish swimming in the undulatory mode is greater than that of the equivalent rigid body. For example, Allen (1961) has shown that the eyes and nares of an oscillating fish may act as roughness elements and cause separation. Allen (1961),

Walters (1962) and Freadman (1979) have shown that gill effluent can also cause boundary layer separation anteriorly.

2. Friction drag is also likely to exceed dead-drag values in a swimming fish, due to boundary layer thinning over those regions of the body producing the propulsive movements (Bone, in Lighthill, 1971).

Not all fish are propelled by the action of undulatory body waves. Many fish employ undulatory and/or paddling median and paired fins (see chapter 7). The results of dead-drag determinations ought to give a good indication of true swimming drag in these cases. Drag coefficients for forms such as the seahorse (*Hippocampus*) and boxfish (*Ostracion*) are several orders of magnitude greater than the dead-drag coefficients of streamlined fish (compare Figs 45 and 46). Low drag coefficients are associated with a low energetic cost of swimming.

Fast cruising fish that capture their prey by rapid pursuit (e.g. Scombriodae) are characterized by low drag coefficients and energetically economical cruising activity. Coral reef fish such as *Ostracion* have high drag coefficients and feed on small, slow-swimming prey (e.g. small crustaceans) and/or plant material. Many coral reef fish are protected from predators by spines and armour and do not flee by employing high speed. These forms are adapted for slow swimming and high manoeuvrability and are not subject to the same morphological constraints as faster swimming, cruising forms. The functional design of streamlined versus unstreamlined fish is discussed further in chapter 7.

Drag reduction

In order to account for the discrepancy between the estimated swimming drag of fish and their rigid-body equivalent drag predicted by the standard hydrodynamic equations of resistance (Gray's Paradox; see chapter 3), biologists and physical scientists began to research possible drag reduction mechanisms in fish. Two main types of drag reduction mechanisms can be recognized: 1. mechanisms designed to maintain a laminar boundary layer; and 2. mechanisms designed to induce a turbulent boundary layer.

At moderate Reynolds Numbers, frictional drag can be kept at a minimum by maintaining a laminar boundary layer over as much of the surface as possible. However, laminar boundary layers are less stable than turbulent boundary layers and are associated with higher pressure drag upon separation. At higher Reynolds Numbers therefore, a turbulent boundary layer may be associated with less drag than a laminar one.

Maintaining a laminar boundary layer

Body form. In the early nineteenth century Sir George Cayley noted the startling similarity in form between the cross-sectional profile of the trout and certain ship hulls. Later von Kármán compared the trout section with several

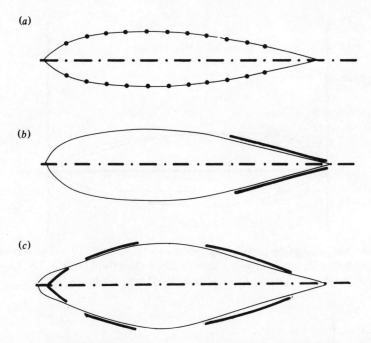

(a)

(b)

(c)

Fig. 48   (a) Sketch of the cross-sectional profile of a modern aerofoil (solid line) compared to that of a trout (solid circles). (b) The profile of a tuna (thin line) compared to that of a laminar drop section (thick line). (c) Profiles of a dolphin (thin line) and of a laminar flow aerofoil (thick line) compared. Based on Hertel (1966).

modern aerofoil profiles (Fig. 48a). Similarly, the cross-sectional profiles of tuna and dolphins can be compared to modern high performance laminar flow profiles (Hertel, 1966; Fig. 48b, c).

The geometric drag-reducing properties of a section are dependent on two parameters: 1. the position of the shoulder (point of maximum thickness); and 2. Fineness Ratio.

The position of the shoulder determines the percentage of the surface area of the section that will experience laminar flow. From the nose to the shoulder, pressure gradients are favourable and boundary layer flow is laminar. Transition is likely to occur at the shoulder and a turbulent boundary layer develops.

A pressure coefficient (p_c) can be defined as

$$p_c = \frac{p - p_0}{\frac{1}{2}\rho U^2} \tag{89}$$

where p is the pressure at a point on the section and p_0 is the ambient static pressure. The variation in p_c over a conventional turbulent flow aerofoil and

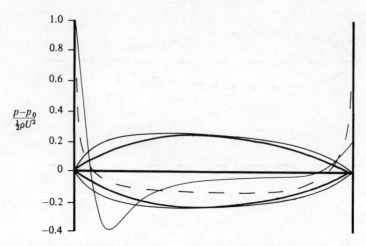

Fig. 49   Variation of the pressure coefficient over the surface of a modern turbulent-flow aerofoil (solid line) and over a laminar flow section (broken line). The Reynolds Number is 1.2×10^5. Data from Goldstein (1938) and von Mises (1959).

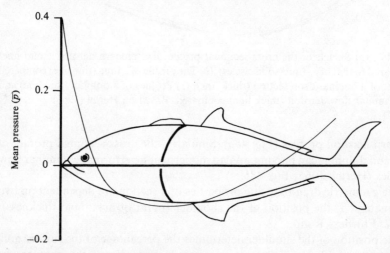

Fig. 50   The pressure distribution over the ventral surface of a dead specimen of *Pomatomus saltatrix*. The contour of maximum thickness over the body is indicated. Modified from Aleyev (1977).

over a laminar profile is plotted in Fig. 49. Fig. 49 shows that (other things being equal): 1. the point of maximum thickness is displaced posteriorly in a laminar profile, relative to its position in a conventional turbulent flow section; and 2. laminar profile aerofoils are characterized by smaller negative pressures at the shoulder and a more even pressure gradient than typical turbulent flow sections.

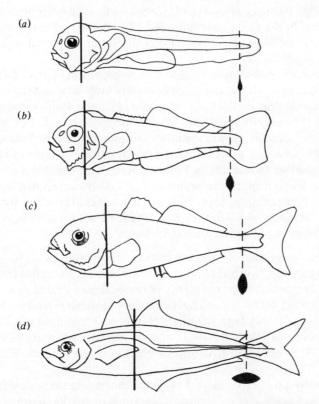

Fig. 51 Variation in the position of the shoulder in the development of *Trachurus mediterraneus*. The development of adipose eye covers and changes in the cross-sectional shape of the caudal peduncle are also shown. Length of fish: (*a*) 0.28 cm; (*b*) 0.68 cm; (*c*) 1.0 cm; (*d*) 43.3 cm. Modified from Burdak (1969).

Houssay (1912) noted that the shoulder is located close to the position of the opercula in most fish. However, in certain scombriods it occurs at 0.6–0.7 of the length back from the snout (e.g. Walters, 1962; Hertel, 1966). Fig. 50 shows that an extended favourable pressure gradient exists over the surface of *Trachurus mediterraneus*. Similar distributions of the pressure coefficient have been found for other scombriods (Aleyev, 1977).

Fig. 51 illustrates the variation in the location of the shoulder in the development of *T. mediterraneus*. With increasing size (and therefore Reynolds Number) the position of the shoulder moves progressively posteriorly. At small size (low Reynolds Number) a laminar boundary layer is maintained as disturbances in the boundary layer are damped by viscous forces. With increasing size and Reynolds Number laminar flow is maintained over most of the fish due to the favourable location of the shoulder. Fig. 51 also illustrates the development of adipose eyelids in *T. mediterraneus*. The adipose eyelids act as fairings. Fairings, grooves and recesses for fins and other protruberances are common in the faster swimming fish. The development of streamlining of the caudal peduncle in its plane of oscillation (see chapter 6) is also shown in Fig. 51.

Fineness Ratio (see chapter 1) affects the development of pressure gradients over the body of a fish and therefore boundary layer flow and the magnitude of the frictional drag on the body. However, Fineness Ratio is more directly related to the magnitude of the pressure drag generated. For maximum volume with minimum surface area the optimum value of the Fineness Ratio is about 4.5. Changes in pressure drag coefficient in streamlined bodies are not very sensitive to changes in Fineness Ratio (see Fig. 11) and a change in Fineness Ratio from the optimum to 3 or 7 is only associated with a 10% increase in pressure drag. Most fusiform fish are characterized by Fineness Ratios in the range of 5.5–7.0 (Hertel, 1966). Values for scombriod fish are typically lower (3.5–5.0; Hertel, 1966; Aleyev, 1977).

Distributed viscous damping. Experiments on the hydrodynamic properties of compliant surfaces, thought to be good analogues of cetacean skin (Kramer, 1960*a*, *b*) are described in chapter 3. Kramer demonstrated that a rubber diaphragm filled with a viscous fluid could act to stabilize laminar boundary layer flow at relatively high Reynolds Numbers. The skin of the Thunnidae and Istiophoridae (Walters, 1962) and of the basking shark (*Cetorhinus*) (Bone, 1974) is similar in structure to the skin of the dolphin described by Kramer.

There is no direct experimental evidence concerning the nature of flow in the boundary layer over cetaceans, scombriods or sharks. Indirect evidence indicates that the boundary layer of cetaceans is mainly turbulent over a range of swimming speeds (Lang & Daybell, 1963). Kramer (1960*b*) found that the frictional drag of the experimental analogue was not reduced in the presence of a fully developed, turbulent boundary layer.

Distributed dynamic damping. Certain trachypterid (e.g. *Desmodema*) and stromateiod (e.g. *Schedophilus*) fish are characterized by a complex skin structure. A system of subdermal canals lies below a keratinized epithelium (Walters, 1963; Bone & Brook, 1973; see Fig. 52). In *Desmodema* the subdermal canals are connected to the surface by pores. Walters (1963) suggests that the canals function to stabilize a laminar boundary layer by distributed dynamic damping. Instabilities in the boundary layer are thought to be damped by fluid sinking into the pores and being transmitted through the canal system to regions of lower pressure. Bone & Brook (1973) suggest a similar function for the subdermal canal system of *Schedophilus*.

There are two unconnected canal systems in *Schedophilus*; one system is in the head region (from nose to shoulder), the other is in the body. It is possible that in the head canal system fluid flows from the high pressure region

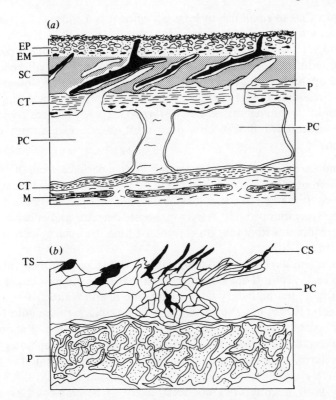

Fig. 52 (*a*) Semi-diagrammatic section of the skin of *Schedophilus medusophagus* (based on Bone & Brook, 1973). (*b*) Skin structure of *Ruvettus pretiosus* (based on a photograph in Bone, 1972). CT, connective tissue layer; EP, external epithelium; EM, external layer of melanophores; M, myotomal muscle fibres; SC, scales and scale pockets; PC, pore canal; P, pore; CS, cycloid scale; p, pigment layer.

about the nose posteriorly to the shoulder and that flow in the main canal system of the body flows in the opposite direction, from the relatively high pressure region of the tail anteriorly towards the shoulder. Ejection of fluid into the boundary layer near the shoulder could play a role in preventing or delaying transition and separation. There is no experimental evidence at this time to support this speculation however. The function of the subdermal canal systems has yet to be investigated experimentally.

Momentum injection into the boundary layer. It has been suggested (e.g. Breder, 1926) that gill effluent ejected into the boundary layer could act to prevent or delay transition and separation. However, experimental evidence suggests that this is probably not the case. Allen (1961) and Freadman (1981) have shown that in small fish at least gill effluent is a source of turbulence, rather than a damping mechanism.

Bainbridge (1961) and Walters (1962) point out that gill effluent could act as a drag reducing mechanism in fish that employ ram-jet gill ventilation where the opercula could function in a manner analogous to slotted wings. Recently Freadman (1981) has shown that ventilatory effluent during cyclic ventilation in swimming striped bass (*Morone saxatilis*) induces turbulence, but that streamlined flow occurs over most of the fish during ram-jet ventilation.

Skin oscillation and boundary layer heating. Active periodic oscillation of the skin has been proposed as a method of boundary layer control (Hertel, 1966; Lang, 1966). It is suggested that local disturbances could be detected by the fish and actively damped out. A very elaborate detector and effector system would be required, with a very small response time. No such system has been identified in any fish.

Cetaceans and some scombriods and lamnid sharks generate heat internally (Carey & Teal, 1969). It has been suggested that these animals could reduce the viscosity of the boundary layer by heating it up (Walters, 1962; Lang, 1966). Webb (1975a) has shown that if a 2 kg skipjack tuna could heat its boundary layer to 30 °C instantly, a frictional drag reduction of about 14% could be expected. However, due to the small contact time of the water (about 0.1 s at preferred swimming speeds) this would require an amount of energy approximately equal to half of the thermal capacity of the fish's body. Boundary layer heating does not seem likely to be employed as a drag reduction mechanism.

Mucus as a drag reducing agent

Breder (1926) thought that mucus could act as a drag-reducing agent by virtue of its 'slippery' nature. However, this violates the 'no-slip condition' (see chapter 1) and is therefore not possible. It has been known for some time that the addition of small amounts of long chain, high relative molecular mass polymers (e.g. polysaccharides) to water can cause a substantial reduction in drag under turbulent flow conditions (e.g. Lumley, 1969). There is also good evidence to show that such substances also have drag reducing properties in pulsed laminar flows (e.g. Hansen, 1973; Kohn, 1973).

Rosen & Cornfield (1971) employed a turbulent flow rheometer to measure the friction drag reduction properties of mucus from a variety of fish. Mucus from the Pacific barracuda gave a 65% friction drag reduction with a 5% mucus–seawater mixture. All of the other fish examined also showed highly significant drag reduction properties. The results of the Rosen & Cornfield experiments are plotted in Figs 53 and 54 for freshwater and seawater fish, respectively.

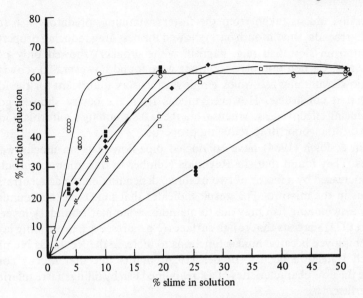

Fig. 53   Friction reduction of mucus from freshwater fish. Symbols refer to speckled rainbow trout (●), kamloops rainbow trout (□), bluegill (△), German brown trout (◆), white crappie (■) and smallmouth bass (○). From Hoyt (1974).

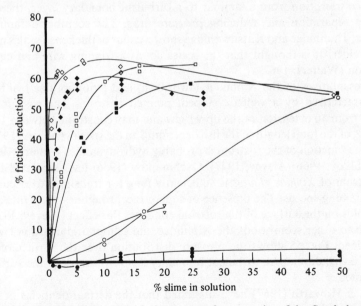

Fig. 54   Friction reduction of the mucus from marine fish. Symbols refer to California bonito (●), sand bass (▽), Pacific mackerel (○), calico kelp bass (■), cowry (□), California halibut (◆) and Pacific barracuda (◇). From Hoyt (1974).

Generally, mucus taken from the faster swimming, predatory fish (e.g. Pacific barracuda, smallmouth bass) showed the best drag reducing properties and that from slow fish (e.g. hagfish, white croaker) showed only poor effectiveness. Certain notable exceptions were found however. The cowry is a very slow-swimming fish which employs mucus as a lubricant for the food traps that it constructs. However, mucus from the cowry showed good drag reducing effectiveness. Mucus from the fast swimming California bonito displayed rather poor drag reducing properties.

Ripkin & Pilch (1964) have conducted pipe-flow tests on mucus–water mixtures. They found that the Reynolds Number to transition to turbulent flow is increased by a factor of two or three, depending on the concentration of mucus in the mixture. This result indicates that mucus in the boundary layer of a swimming fish may act to maintain a laminar boundary layer.

Hoyt (1974) suggests that rough surfaces (e.g. nares, eyes, protruding jaws) can be employed because mucus functions to increase the Reynolds Number at which transition occurs. Breder (1926) points out that mucus could improve the flow characteristics of a streamlined fish by filling in irregularities in its surface.

Inducing and maintaining a turbulent boundary layer

Prominent roughness elements (e.g. spines, rough scales) are thought to act to induce transition from a laminar to a turbulent boundary layer, thereby delaying separation and reducing pressure drag. The scombriod families Cybidae, Thunnidae and Katsuwonidae show a collar of thickened scales near the shoulder. It is thought that the scales act as a tripping wire and cause transition (Walters, 1962).

Representatives of the Xiphiidae (swordfish) and Istiophoridae (sailfish) are characterized by a well developed, pointed rostrum. The rostrum is formed from an extension of the upper jaw and may constitute between 15% and 45% of the total length of the fish depending on the species (Aleyev, 1977). The basic function of the rostrum is probably hydrodynamic (Walters, 1962; Ovchinnikov, 1966; Aleyev, 1977). Ovchinnikov (1966) has calculated that the rostrum of *Xiphias gladius* is sufficiently long for transition to occur at low swimming speeds. The presence of supercritical roughness elements and small folds on the surface of the rostrum enhances this effect (Aleyev, 1977). Relative to other scombriods the Xiphiidae and Istiophoridae are not well streamlined. Fig. 55 shows the pressure distribution over the surface of *X. gladius*, a well developed, turbulent boundary layer over most of the surface is indicated.

Bone & Howarth (1967) have suggested that the dermal denticles of the dogfish probably act as roughness elements and induce transition. Ctenoid scales probably function in a similar fashion. Burdak (1969) describes the development of scales in *Mugil saliens*. In juvenile fish, scales are absent.

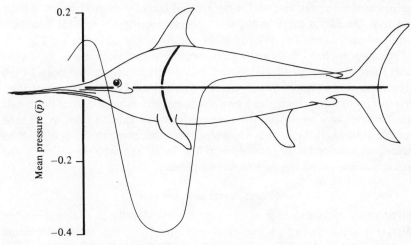

Fig. 55   The pressure distribution over the ventral surface of a dead specimen of *Xiphias gladius*. Modified from Aleyev (1977).

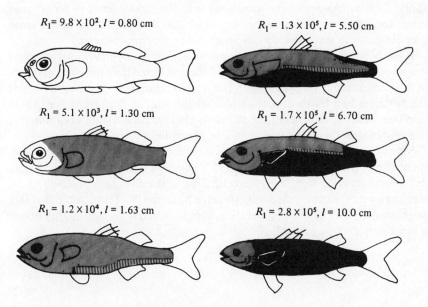

Fig. 56   Development of scales in *Mugil saliens*. □, no scales; ▥ cycloid scales; ▦ cycloid and ctenoid scales; ■ ctenoid scales. After Aleyev (1977).

Cycloid scales appear first and these are followed by ctenoid scales. By the time that the fish is large enough to approach critical Reynolds Numbers during swimming most of its body is covered with ctenoid scales (see Fig. 56).

The castor oil fish (*Ruvettus*) is also covered with ctenoid scales. Below the scales there is a well developed system of subdermal canals that open to the surface via pores (Bone, 1972; see Fig. 52*b*). Bone (1974) suggests that as *Ruvettus* oscillates its body during swimming, water is ejected from the canals into the boundary layer from the convex leading surface and drawn in at the concave trailing surface. It is suggested that this mechanism could act to prevent separation. It is possible that the large overlapping scales of the Osteoglossidae could function in the same way.

Finlets, keels and fairings

The presence of fairings (e.g. adipose eye covers, slots for the paired fins) and transverse streamlining of the caudal peduncle in the faster swimming Scombriodae has already been noted. Many of the same fish are characterized by a series of small finlets on the dorsal and ventral surfaces of the caudal peduncle (see Fig. 26*a*). Harris (1950) noted similar structures in the fossil shark *Diademodus hydei* and suggested that the finlets acted to rectify and direct the flow over the caudal peduncle (see also Walters, 1962). In this sense they function in the same way as wind tunnel damping screens.

It has been suggested that the adipose fin and small second dorsal and anal fins of certain salmonids and characids function in the the same way as finlets in the Scombriodae (Aleyev, 1977). In certain Carangidae (e.g. *Selene vomer*) the posterior two-thirds of both the dorsal and anal fin flaps from side to side in phase with the movements of the body (Blake, personal observations). It is possible that these fins play a role in controlling cross-flows.

Many Scombriodae, Istiophoridae and some sharks have paired keels on each side of the caudal fin. Aleyev (1963) suggests that the keels function as deflectors, ordering flow over the caudal fin and reducing induced drag. In some cases (e.g. *Lamna*) the keels converge posteriorly. It is possible that this might reduce the chance of separation over the caudal fin by bringing about a local acceleration of the flow.

5

Undulatory propulsion

Introduction

This chapter is concerned with fish that are propelled by the action of undulatory body waves. Many of the forms discussed are common pelagic, fusiform fish that are characterized by the presence of a distinct caudal fin. In forms such as the eel (*Anguilla*) the waveforms that pass down the body are well developed and obvious to the unaided eye. At the other extreme there are fish in which the wavelength of the body waves is long relative to body length, and only the rapid oscillation of the caudal fin can be seen (e.g. Scombriodae). We begin by classifying the various undulatory modes of propulsion on the basis of kinematic parameters that can be readily determined (e.g. wavelength, amplitude). It is important to realize that the classification is arbitrary and serves merely as a convenient division of the observed kinematic variation.

Four modes are defined: anguilliform (the pure undulatory mode as seen in the eel), subcarangiform (movements typical of the cyprinids, e.g. goldfish), carangiform (referring to the more restricted waveforms seen in the Carangidae), and thunniform (as seen in the scombriods and their relatives). Most experimental work has focused on subcarangiform swimming, and a discussion of kinematics in relation to swimming speed and size concentrates on subcarangiform swimmers.

In the following section the kinematics of unsteady swimming behaviour is discussed under the headings of fast-starts and turning. The time course and kinematic stages of fast-starts and turns are discussed. Variation in the fast-start behaviour and performance of a variety of fish is considered. Much of the work discussed in this section is relatively recent. There is still a great deal of scope for experimental and theoretical work in this area.

A sound knowledge of the kinematics of swimming is necessary for the formulation of hydromechanical models from which such quantities as thrust, power and efficiency can be calculated. The earliest hydromechanical models deal with steady swimming and consider the localized resistive forces that are generated as a short segment of the fish interacts with the water (e.g. Taylor,

71

1952). The magnitude of the resistive forces is assumed to be dependent on the instantaneous relative velocity and angle of attack of the body, and in this sense the model may be regarded as quasi-static.

More recently, hydrodynamicists have developed reactive theories of swimming based on the inertial forces generated by the propulsive surfaces of the fish in a perfect fluid (e.g. Lighthill, 1960, 1970, 1971; Wu, 1961, 1971a, b, c, d). The inertial forces are dependent on the rate of change of the relative velocity of the propulsive parts and on the mass of water which they displace.

Lighthill's reactive theory of swimming is referred to as elongated-body theory. The theory is considered in detail, and in a simplified form (bulk-momentum model). Thrust forces calculated from elongated-body theory for subcarangiform swimmers are compared with the frictional equivalent rigid-body drag, calculated from the standard hydrodynamic equations of resistance. Drag coefficients calculated from elongated-body theory are five to ten times greater than the rigid-body values. The discrepancy may be due to the effects of boundary layer thinning in the swimming fish.

The concept of efficiency as a measure of performance is discussed next. Values of propulsive efficiency calculated from elongated-body theory are compared with experimentally determined values for subcarangiform swimmers. Good agreement is found at higher cruising speeds where values of propulsive efficiency exceed 80%. Of the metabolic energy available for swimming, only about 20% is delivered by the propulsive musculature as mechanical work (see chapter 2). The product of propulsive and muscle efficiency defines the overall aerobic efficiency of the system. Values of aerobic efficiency range from about 15 to 30%.

Elongated-body theory has been developed in recent years to cover the unsteady swimming behaviour of fish (Weihs, 1972, 1973a). The modified form of the theory contains terms that take account of the vorticity shed by sharp edges such as fins. The theory can be employed to calculate the forces and moments associated with fast-starts and turns. Calculated turning radii compare well with observational data.

Two hydromechanical theories of thunniform swimming are discussed, the first is a resistive theory due to von Holst & von Kuchemann (1942), the second a reactive, two-dimensional theory developed by Lighthill (1969, 1970). Parry (1949) employed the von Holst & von Kuchemann theory to calculate swimming thrust in a whale in relation to tail-beat frequency and swimming speed. Thrust values for the whale are compared to equivalent rigid-body drag based on either a fully developed laminar or turbulent boundary layer. Lighthill's reactive theory of thunniform swimming is employed to calculate thrust coefficients and propulsive efficiency. Calculations show that the thrust coefficient and propulsive efficiency are inversely related.

Lastly, kinematics and hydromechanics are discussed in relation to functional morphological design. The influence of a variety of morphological features on steady and unsteady swimming performance is considered. Good steady swimming performance in the anguilliform and subcarangiform modes requires a large depth of section anteriorly over the centre of mass, a relatively inflexible anterior section and a narrow caudal peduncle. Many of the morphological requirements for good unsteady swimming performance are not compatible with these features. For example, a deep caudal peduncle and flexible body enhance fast-start performance. The morphological design of many fish can be interpreted as a compromise between those features required for steady swimming and those associated with good unsteady swimming performance. Only the scombriods are ideally designed for efficient cruising.

Kinematics of steady forward swimming

Classification of modes

The propulsive movements of fish were first classified by Breder (1926). Among those forms that are propelled by the undulatory action of the body and/or caudal fin, three main modes were distinguished: anguilliform (named after the eel, *Anguilla*), carangiform (named after *Caranx*) and ostraciiform (named after the boxfish *Ostracion*).

In the anguilliform mode the whole body is bent into backward moving waves. The specific wavelength (wavelength of the body waves divided by body length) is less than one. At least one half of a wavelength is present on the body at any given time. Often more than a complete wavelength can be seen. The amplitude of the waves increases posteriorly and is large over most of the length of the body. Typically, anguilliform swimmers are long, thin fish with a cylindrical anterior section and a laterally compressed posterior section (e.g. Anguillidae, Petromyzontidae, Ammodytidae). If present, the caudal fin is of low aspect ratio (e.g. Siluriformes). Anguilliform movements in the eel and in the dogfish are illustrated in Figs 57 and 58, respectively.

Many common pelagic, fusiform fish swim in the subcarangiform mode (e.g. *Salmo*, *Carassius*, *Leuciscus*). The specific wavelength is less than one. More than one half of a wavelength, but rarely more than one complete wavelength may be present on the body at any given time. The amplitude of the waveforms increases rapidly over the posterior third to half of the body length. A caudal fin of moderate aspect ratio is typical of subcarangiform swimmers. Subcarangiform swimming in the whiting (*Gadus merlangus*) is illustrated in Fig. 59.

In the carangiform mode proper, specific wavelength may be less than one and up to one-half of a wavelength may be seen on the body at any instant. Undulations are confined to the posterior third or so of the body. Amplitude increases rapidly in this area. Most Carangidae, Characidae and Mormyridae swim in the carangiform mode.

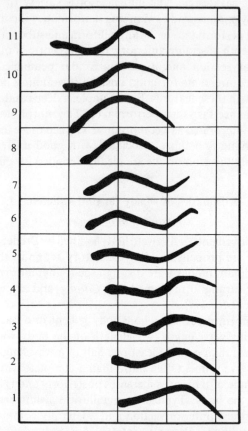

Fig. 57   Anguilliform swimming in the eel (*Anguilla vulgaris*, $l = 0.07$ m). The interval between each position is 0.09 s. Modified from Gray (1968).

The Scombriodae, Istiophoridae, Xiphiidae and Lamnidae are characterized by a narrow caudal peduncle and a large, half moon shaped caudal fin (lunate tail) of high aspect ratio. Undulations of the body are confined to the caudal peduncle and fin. The term thunniform (after the tunnyfishes) is employed to describe the movements of these fish.

Most ostraciiform swimmers are unstreamlined and encased in a bony armour (e.g. Ostraciidae). A rigid, low aspect ratio caudal fin pivots about the base of the caudal peduncle. Ostraciiform propulsion is discussed in detail in chapter 6.

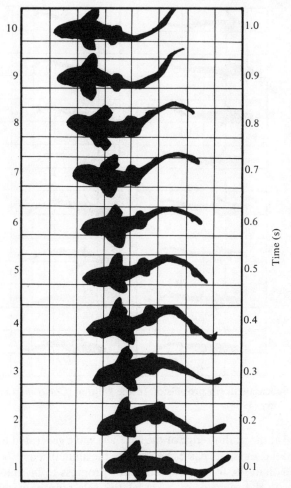

Fig. 58 Anguilliform propulsion in the dogfish. The animal is about 0.35 m in length and the interval between successive positions is 0.1 s. Modified from Gray (1968).

Anguilliform kinematics

The first vigorous analysis of anguilliform swimming was performed by Gray (1933*a*, *b*, *c*). Much of the account that follows is based on Gray's work on eel swimming (Gray, 1933*a*).

The entire body is thrown laterally into a propulsive wave as a result of the phased contraction of the body muscles. It is convenient to view the body as being composed of a series of small segments. Each segment moves transversely with a periodic motion. Posterior segments lag behind anterior segments. The amplitude of the motion increases towards a maximum which

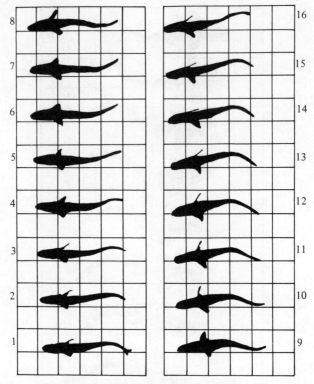

Fig. 59   Subcarangiform propulsion in the whiting (*Gadus merlangus*, $l = 0.30$ m). Based on Gray (1968).

is reached at the trailing edge of the fish. The wave generated by the transverse motion of the segments passes back down the animal at a velocity V that must be greater than the forward speed U for progress to be made.

   The locus of a segment in space relative to a point moving along the axis of progression (A, B) at U (i.e. relative to the fish) is a transverse figure of eight (see Fig. 60). The figure of eight is more marked posteriorly where the amplitude of the transverse motions is large relative to the wavelength of the body wave. Relative to a fixed point (i.e. relative to a stationary observer) each segment describes a sine curve of wavelength λ_s. The value of λ_s is about the same for all segments and is less than the wavelength of the body, λ_b (Fig. 61). Although λ_s is approximately constant for all portions of the body, the amplitude of the movements increases posteriorly.

   Each segment subtends an angle (θ_s) between the track of the fish and the transverse axis of movement of the segments (C, D). The value of θ_s decreases posteriorly as the amplitude of λ_s increases. The angle made by any segment of the body and the axis C, D (θ_b) also decreases from head to tail. Since λ_b

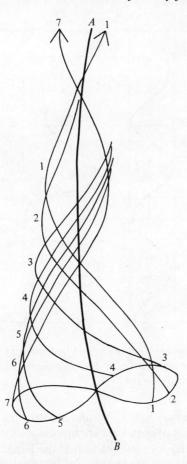

Fig. 60 Diagram to show the transverse figure-of-eight motion of the trailing edge segment in anguilliform swimming. Successive positions of the body waves are shown corresponding to tail positions 1 to 7. The line A, B is the axis of progression. Based on Gray (1968).

is greater than λ_s, it follows that θ_b must be greater than θ_s. The angle of attack (α) of a segment is given by the difference between θ_b and θ_s (i.e., $\alpha = \theta_b - \theta_s$).

The transverse velocity of a segment (W) is zero at the positions of maximum displacement and at a maximum as the segment crosses A, B. For about half of the amplitude of the displacement W is close to its maximum value however. The resultant velocity of a segment (W_r) is given by

$$W_r = (W^2 + U^2)^{\frac{1}{2}} \qquad (90)$$

The relationship between θ_s, θ_b, α, W, U and W_r is shown diagrammatically

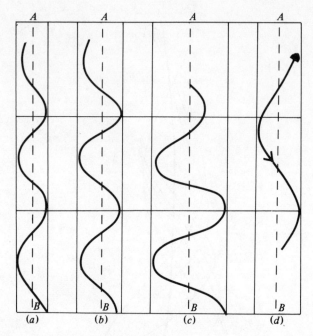

Fig. 61   The path of segments in space traced by the head (*a*), midpoint of the body (*b*) and the trailing-edge segment (*c*) in *Anguilla*. The wavelength of the body wave is indicated in (*d*). Based on Gray (1933*a*).

in Fig. 62. Fig. 62 shows that: 1. W_r and α also vary with position on the body; and 2. the values of α and W_r are inversely related, with α decreasing towards the tail and W_r decreasing towards the head.

For forward thrust to be generated α must be positive. This occurs when θ_b and λ_b exceeds θ_s and λ_s, which is the case when V is greater than U.

Subcarangiform kinematics

There is little difference between the basic body movements of typical pelagic fish such as the trout and those of the eel. Gray (1933*c*) noted that the main difference between the propulsive movements of the eel and the whiting is that in the latter the amplitude of the body waves remains small until the waves reach the posterior of the body where a rapid increase in amplitude occurs. Wardle & Videler (1980) have studied swimming in the cod (*Gadus morhua*) which swims in the subcarangiform mode. The body movements of *G. morhua* during steady forward swimming are shown in Fig. 63.

The most complete study of subcarangiform swimming kinematics was performed by Bainbridge (1963) on trout (*Salmo gairdneri*), dace (*Leuciscus leuciscus*) and goldfish (*Carassius auratus*). The account that follows is drawn

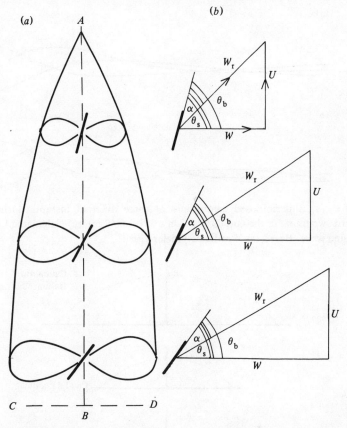

Fig. 62   (*a*) Schematic diagram showing the motion of successive segments back from the nose between positions of maximum displacement. (*b*) Diagrams indicating the variation in the principal kinematic parameters for successive segments. Based on Gray (1933*a*, 1968).

mainly from Bainbridge (1963). Fig. 64 shows the variation in the amplitude and transverse velocity of different segments along the body of a specimen of *Leuciscus* during steady forward swimming. We note that:

1. The body movements are symmetrical about the longitudinal axis of the fish.

2. Amplitude and transverse velocity decrease from the nose to the operculum where both are at a minimum.

3. From the operculum to the trailing edge of the caudal fin, amplitude and transverse velocity increase. Most of the increase occurs between the trailing edge of the dorsal fin and the trailing edge of the caudal fin.

The form of the curves obtained from plots of transverse velocity against

(a)

(b)

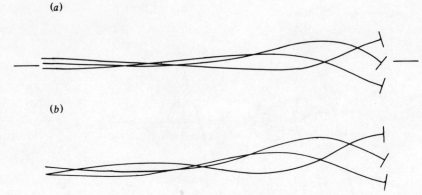

Fig. 63   (a) Superimposed centre-lines at three different instants during steady forward swimming in the cod ($U = 0.7$ m s^{-1}, $l = 0.42$ m). (b) Calculated lines based on a sine wave. Based on Wardle & Videler (1980).

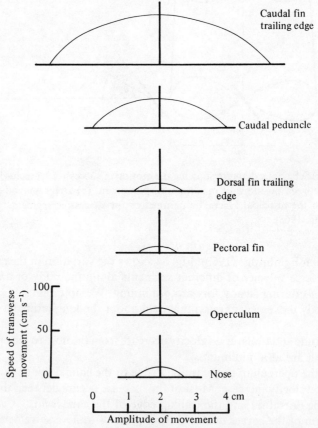

Fig. 64   Transverse velocity and amplitude of different segments along the body of *Leuciscus leuciscus* ($U = 0.48$ m s^{-1}). Modified from Bainbridge (1963).

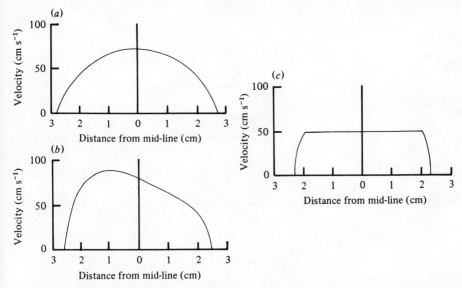

Fig. 65 Diagrammatic representation of the lateral velocity of a trailing edge segment of a dace, showing a symmetrical (*a*), asymmetrical (*b*) and 'flat-topped' (*c*) velocity distribution. Based on Bainbridge (1963).

distance from the midline exhibited variation. Symmetrical curves were typical of steady forward swimming (Fig. 65*a*). Fig. 65*b* shows an asymmetrical curve in which the maximum value of *W* occurs early in the tail-beat cycle. Bainbridge described other examples in which the maximum values of *W* were obtained late in the cycle. Asymmetrical curves were often associated with small accelerations and decelerations. Similar tail-beat patterns are described by Freadman (1981) for striped bass (*Morone saxatilis*) and bluefish (*Pomatomus saltatrix*). In Fig. 65*c* the value of *W* is shown rising rapidly at the beginning of the stroke and then plateauing, remaining constant over most of the cycle.

Fish that swim in the subcarangiform mode are characterized by a well-developed, highly flexible caudal fin. The behaviour of the caudal fin during swimming is determined by its structure, the nature of the propulsive wave that is transmitted to it from the body and, possibly, by a small amount of intrinsic musculature. These three factors interact to cause bending (dorso–ventral and anterio–posterior) and changes in the area of the fin during the tail-beat cycle.

Bainbridge (1963) found that for *Leuciscus leuciscus* during steady forward swimming, the minimum and maximum amounts of dorso–ventral bending of the caudal fin occurred when the fin was approximately one-quarter and three-quarters of the way through the stroke, respectively. A consequence of

(a)

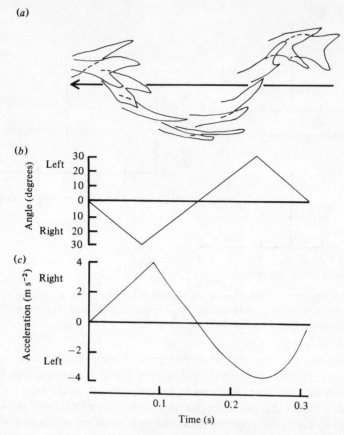

Fig. 66   (a) Path of the caudal fin of a bluefish (*Pomatomus saltatrix*, $l = 0.48$ m, $U = 0.87$ m s^{-1}). (b) Variation in the angle of attack of the tail fin. (c) Lateral acceleration of the tail fin. Data from Du Bois & Ogilvy (1978).

the dorso–ventral bending of the caudal fin is that the dorsal and ventral tips of the fin reach the points of maximum displacement before the centre of the fin.

Anterio–posterior bending of the caudal fin occurs in the horizontal plane (i.e. 90° out of phase with the dorso–ventral bending of the fin). In consequence of the anterio–posterior bending, each segment of the fin presents an angle of incidence to the axis of progression of the fish. Bainbridge (1963) measured mean angles (for the entire fin over the entire fin-beat cycle) of about 15° for goldfish and dace. Du Bois & Ogilvy (1978) have made similar measurements on bluefish. The path of the caudal fin of a specimen of *P. saltatrix* during steady forward swimming is shown diagrammatically in Fig. 66*a*. The variation of the angle of the tail fin to the axis of progression is shown in Fig. 66*b*. Similar results were obtained for *G. morhua* by Videler & Wardle (1978).

McCutcheon (1970) has pointed out that the tail fin is not simply a sheet of compliant material. The fin rays are composed of bony half rays, each of which behaves like a flexible girder. The half rays are free to move relative to each other up to a limit. Anterio–posterior bending is determined mainly by hydrodynamic forces until the limit is reached. McCutcheon views the caudal fin as a self-cambering hydrofoil, with the fin automatically adjusting its camber, so that the hydrodynamic load is distributed the same way at all swimming speeds. Videler (1975) provides a detailed description of the morphology and movement of the tail in *Tilapia nilotica*. Videler emphasizes the likely role of the skin in transmitting bending moments generated by the body to the tail.

Bainbridge (1963) measured changes in the area of the caudal fin over the tail-beat cycle in several species of freshwater fish. He noted that the span of the tail fin fluctuated symmetrically about the axis of progression and that the fin span and area were not directly correlated due to the influence of dorso–ventral bending. The caudal fin presents its smallest area at the extremes of the stroke. As the fin accelerates towards the axis of progression its area increases. Maximum area is projected when the fin moves at or close to its maximum transverse velocity. According to Bainbridge the observed changes in area of the caudal fin during the fin-beat cycle allow for: (1) a rapid initial acceleration of the fin, due to its small area at the beginning of the stroke; (2) a high and uniform thrust when transverse velocity is falling by increasing area at that time; (3) slowing of the fin by its increased area; and (4) minimum drag by presenting a small area when at the extremes of the stroke.

Carangiform kinematics

Few observations have been made on fish that swim in the carangiform mode. Essentially, the basic kinematics of carangiform swimming are not very different from those typical of subcarangiform swimmers. Webb (1975*a*) points out that the caudal fin of carangiform swimmers tends to be rigid and exhibit a 'swept-back' planform. The stiffness of the caudal fin in forms such as the mackerel (*Scomber*) probably reduces anterio–posterior bending substantially. Dorso–ventral bending of the caudal fin is probably also reduced by this stiffness. Lindsey (1978) suggests that it is likely that the caudal fin of carangiform swimmers is under less control by the caudal-fin intrinsic musculature than is the case with subcarangiform fish. The caudal-fin intrinsic musculature is only poorly developed in *Clupea* and *Caranx*.

Lighthill (1969) points out that the rigid caudal fin of carangiform swimmers is not simply wagged from side to side about the caudal peduncle.

The angle of inclination of the fin is changed as it moves from one extreme of the stroke to the other so that the fin always generates a backward-facing component even when moving away from the axis of progression.

Thunniform kinematics

Fish that swim in the so-called thunniform mode are characterized by a distinctive, high aspect ratio, lunate caudal fin. The lunate tail has evolved independently in four major groups of aquatic vertebrates: the elasmobranchs, percomorph fish, cetaceans and reptiles. The elasmobranchs are represented by the lamnid sharks. Among the percomorphs (the Percomorphi is a large order of teleosts which contains 120 families), the Scombriodae (e.g. tunnies, bonitos), Luvaridae (e.g. louvars), Istiophoridae (e.g. sailfish, marlin), Xiphiidae (swordfish), Coryphaenidae (dolphinfishes) and Stromateidae (butterfishes) swim in the thunniform mode. The extinct reptile *Ichthyosaurus* possessed a very well developed lunate tail and probably swam in the thunniform mode.

Thunniform swimmers have attracted a great deal of attention from biologists and physical scientists in recent years, largely because of the high speeds they attain (see chapter 3). The thunniform swimming mode is not always associated with high speed however. The deep-sea trichurid *Aphanopus carbo* swims in the anguilliform mode using a body wave. When stalking its prey, however, *Aphanopus* swims slowly, employing its high aspect ratio caudal fin (Bone, 1971).

Propulsive movements are of long wavelength (specific wavelength may exceed 1.0) and confined to the caudal peduncle and fin. In the percomorphs the body muscles deliver the propulsive force to the tail by a system of tendons which run over a double-joint system (Nursall, 1958; Slijper, 1958; Kramer, 1960a; Fierstine & Walters, 1968). The first joint is located at the base of the caudal peduncle and facilitates lateral bending of the caudal fin. The second joint is at the posterior of the caudal peduncle and allows for a rotational motion of the caudal fin. This enables the caudal fin to change its angle of attack over the fin-beat cycle.

Fig. 67 illustrates a cycle of propulsive movements in *Euthynnus affinis* which swims in the thuniiform mode. Fig. 67e shows that the motion of the caudal fin is similar to that of a trailing edge segment of a subcarangiform swimmer. However, anterior to the caudal fin there is very little lateral movement of the body. Maximum amplitudes are about $0.2l$ and do not vary with length (Hunter & Zweifel, 1971). Fierstine & Walters (1968) have measured W and α in *E. affinis*. Values were similar to those recorded by Bainbridge (1963) for fish that swim in the subcarangiform mode.

Movements of the caudal fin are relatively simple (see Fig. 67e) as little bending of the caudal fin occurs due to its high stiffness. However, some dorso–ventral bending occurs in *E. affinis*. The centre of the caudal fin is more

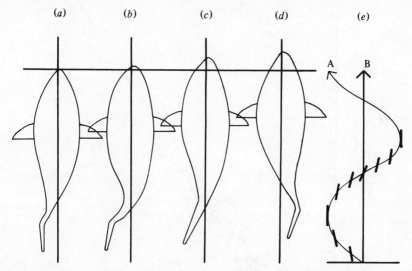

Fig. 67   (*a–d*) Successive body positions of *Euthynnus affinis*, swimming in the thunniform mode. (*e*) Diagram of the path of the caudal fin. Adapted from Fierstine & Walters (1968).

rigid than the tips in most thunniform swimmers and in consequence leads in the oscillation.

In cetaceans the tail fin (flukes) oscillates in a vertical plane, and there are other differences between the kinematics of lunate tail swimming in cetaceans and percomorphs. For example, bending of the posterior part of the body is more marked in cetaceans than it is in percomorphs (Slijper, 1958, 1961; Lang & Daybell, 1963). In *Tursiops truncatus* the tail moves up and down about a centre of rotation that is just posterior to the dorsal fin, throwing the posterior third or so of the body into a sharp curve (Parry, 1949). Parry (1949) describes the tail-beat cycle in *T. truncatus* during steady forward swimming. When the tail is at the top of the stroke the flukes are horizontal, and therefore present little resistance to forward motion. As the tail moves towards the axis of progression the flukes bend upward. They reach their maximum angle relative to the horizontal as the axis of progression is crossed. The angle of the flukes returns to zero by the end of the stroke. During the second half of the cycle the movement is reversed.

Kinematics, speed and size

From the results of a study on the relations between size, speed and swimming kinematics in trout, goldfish and dace, Bainbridge (1958) concludes that:
1. Speed is directly proportional to length for any given tail-beat frequency.

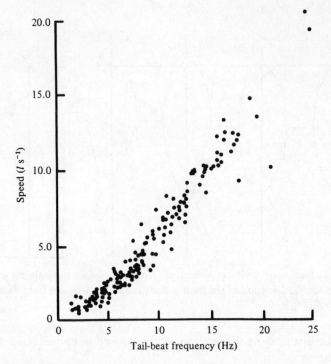

Fig. 68   Relation between tail-beat frequency and speed in the dace (*Leuciscus leuciscus*). Based on Bainbridge (1963).

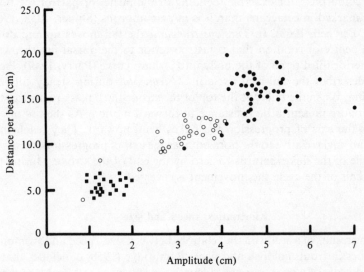

Fig. 69   Relation between distance travelled per unit tail beat and amplitude for dace: (●), 0.24 m; (○), 0.17 m and (■), 0.09 m. From Bainbridge (1958).

2. Above a tail-beat frequency of about 5 Hz speed is directly proportional to tail-beat frequency (see Fig. 68).

3. The distance travelled per tail beat is directly dependent on the amplitude of the tail beat (see Fig. 69).

4. Amplitude increases with increasing frequency up to tail-beat frequencies of about 5 Hz.

5. The value of the specific maximum amplitude (maximum amplitude divided by body length) is independent of size.

6. Maximum frequency of oscillation and size are inversely related.

Bainbridge found that swimming speed could be predicted from the equation

$$U = \tfrac{1}{4}(l(3f-4)) \tag{91}$$

where f is the frequency of the propulsive waveforms. Tail-beat frequency is plotted against specific swimming speed for dace in Fig. 69. Webb (1971a) gives similar results for trout. Noting that the propulsive wave speed must exceed the forward speed and that $V = f\lambda_b$, we can write

$$f = \frac{4(V+l)}{3l} \tag{92}$$

$$\frac{\lambda_b}{l} = \frac{3U_{max}}{4(V_{max}+l)} \tag{93}$$

Assuming that the maximum swimming speed is $10\ l\ s^{-1}$, λ_b/l will be 0.68 and $\lambda_b = 0.681$ (Webb, 1971a). For $U_{max} = 20\ l\ s^{-1}$, λ_b increases to $0.71l$. Webb found good agreement between values of λ_b predicted from eqn 93 and experimental values. According to eqn 93, for a given value of U_{max}, λ_b/l should be a constant, and so for a fish of given length λ_b should not vary with speed. Videler & Wardle (1978) found this to be the case for cod.

Webb (1971a, 1973a) confirms Bainbridge's observations on the relation of f and specific amplitude to speed and shows that the product of the two is linearly related to speed. Both authors found variation in frequency and amplitude at low speeds. However, this is not the case for all fish. Hunter & Zweifel (1971) found that f could be related to U over the whole speed range for a variety of fish (i.e. $U/l = kf + b$, see Fig. 70) and that swimming speed could be accurately predicted from

$$U - U_{min} = l(kf - f_{min}) \tag{94}$$

where U_{min} and f_{min} are the length-dependent, minimum swimming speed and tail-beat frequency, respectively. In certain scombriods, amplitude increases with swimming speed and the wavelength of the propulsive wave decreases at higher speeds (Magnuson & Prescott, 1966; Pyatetskiy, 1970).

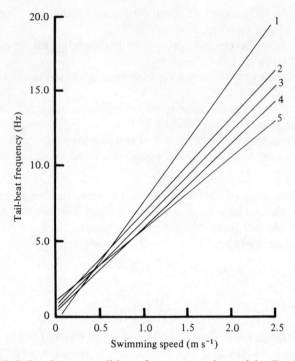

Fig. 70 Relation between tail-beat frequency and speed in *Trachurus, Scomber, Leuciscus, Carassius* and *Salmo* (curves 1–5, respectively). Based on Hunter & Zweifel (1971).

Kinematics of unsteady swimming

Fast-starts

The acceleration performance of fish is striking. Accelerations of the order of 40–50 m s^{-2} have been recorded for a variety of fish ranging from trout to tuna (e.g. Gero, 1952; Bainbridge, 1958; Hertel, 1966; Fierstine & Walters, 1968; Weihs, 1973*a*; Webb, 1977, 1978, 1981). Fig. 71 shows a typical lunging fast-start in a trout. On the basis of this type of information Weihs (1973*a*) identified three stages in the fast-start behaviour of trout and pike:

1. The preparatory stage (kinematic stage one): The fish assumes an L-shaped posture by moving its tail laterally until it is at right angles to the main body axis. Due to the sideways movement of the tail, yawing forces are generated resulting in some rotation of the head.

2. The propulsive stroke (kinematic stage two): The caudal fin moves sideways rapidly so that at the end of its stroke it lies perpendicular to the main body axis at the other side of the midline. Yawing forces cause a large rotation of the head and the fish moves forward at an angle to the plane of

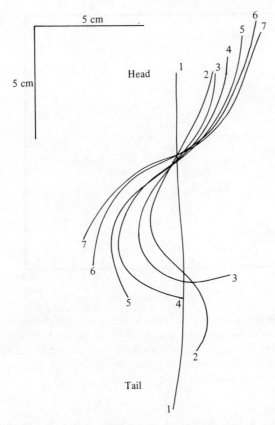

Fig. 71 Tracings of the centre-line of a trout ($l = 0.33$ m) during a fast-start. Successive positions are at intervals of 0.025 s. From Weihs (1973*a*).

its original position. The propulsive lateral movements of the caudal fin may be repeated several times.

3. The final stage (kinematic stage three): The fish may continue to move forward propelled by normal propulsive movements or simply glide.

Webb (1976) identified two fast-start kinematic patterns in the trout: the L-type start described by Weihs; and an S-type start in which the body is bent into a double flexure during kinematic stages one and two. Fast-starts of the S-type are not usually associated with large recoil movements of the body. Webb found that S-type fast-starts are more common in larger fish. For both types of fast-start behaviour, acceleration rates and the distance covered and velocity attained after a given time for fish accelerating from rest are independent of size (Webb, 1976). Maximum acceleration rates of about 35 m s^{-2} were recorded.

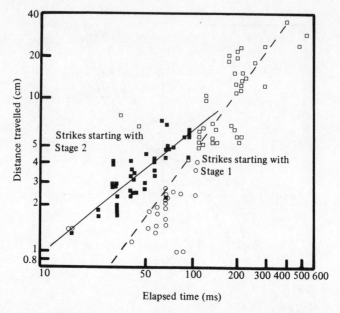

Fig. 72   Relation between distance travelled and elapsed time during fast-starts in *Esox*. Square symbols show distances and times from the start of acceleration to contact time with prey (minnow). Circular symbols show distances and times to the end of stage one. Solid squares represent pattern B strikes, open squares pattern A strikes. Based on Webb & Skadsen (1980).

Pike (*Esox*) show S-type fast-start behaviour. Webb & Skadsen (1980) identified two distinct patterns of behaviour within this category:
Pattern A: Strikes commence from a stretched straight position and show the normal three-stage kinematic sequence.
Pattern B: The first kinematic stage is omitted.
Pattern B-type S-starts are usually employed for close-range strikes at prey. The relation between distance travelled and time for pattern A and B fast-starts is shown in Fig. 72.

Webb (1978) studied the fast-start performance of seven different teleost species (*Etheostoma caeruleum, Cottus cognatus, Notropis cornutus, Lepomis macrochirus, Perca flavescens, Salmo gairdneri* and *Esox*). Normal three-stage kinematics were found for all seven species. Webb found that:
1. Acceleration rates are not a function of size. Maximum acceleration rates of 22.7–39.5 m s^{-2} and mean rates of 6.1–12.3 m s^{-2} (averaged to the completion of kinematic stage two) were found.
2. Maximum velocities attained and distances covered in each fast-start stage

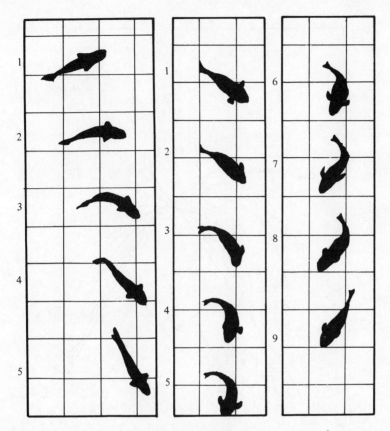

Fig. 73   Turning movements in the goldfish (*Carassius auratus,* $l = 0.15$ m). The interval between frames is 0.04 s. Adapted from Gray (1933*a*).

varied and were related to length. Times to the end of kinematic stage one (t_1) and two (t_2) can be described by:

$$t_1 = 0.0019l + 0.026 \tag{95}$$
$$t_2 = 0.0035l + 0.043 \tag{96}$$

The distance relations can best be described by power functions:

$$s_1 = 0.18l^{0.94} \tag{97}$$
$$s_2 = 0.38l^{1.01} \tag{98}$$

where s_1 and s_2 are the distances covered during kinematic stages one and two, respectively.

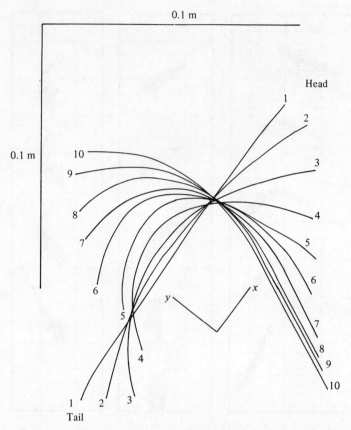

Fig. 74   Tracings of the centre-line of a goldfish during a 'standing start' turn. The interval between frames is 0.04 s. From Weihs (1972).

Turns

Weihs (1972) describes the kinematics of turning in the goldfish (*Cyprinus auratus*) and the rudd (*Cyprinus erythrophalmus*). His descriptions are based upon film records made by Gray (1933c, 1968). Weihs identified three stages in the 'standing start' turn of the goldfish (see Fig. 73):

1. Stage one: The head rotates into the direction of the turn. This causes momentum changes which result in a net force opposing the rotation. At the same time the tail rotates in the opposite direction (see Fig. 74, positions 1–4). There having been no initial tangential velocity, the net moment in the direction of the turn is small. At the end of stage one the fish is curved with its head close to the final direction of travel.

2. Stage two: A large increase in the curvature of the central part of the body occurs. The tail changes its direction causing a change in the force balance,

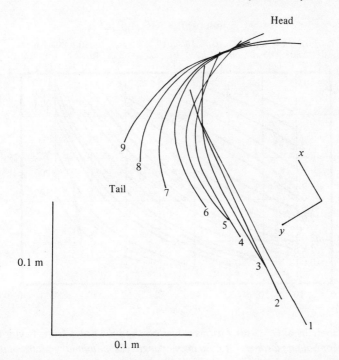

Fig. 75 Tracings of the centre-line of a rudd during a turn. The interval between frames is 0.04 s. From Weihs (1972).

resulting in a net force in the direction of the turn (see Fig. 74, positions 5–7).
3. Stage three: The tail straightens out (Fig. 74, positions 7–10).

The rudd analysed by Weihs was travelling forward at 0.6 m s^{-1} when a turn was initiated. However, the same three kinematic stages can be identified (Fig. 75). The main difference between the 'standing start' turn of the goldfish and the turn of the rudd is that the amplitudes of the lateral movements of the tail of the rudd during stage one are far smaller than those of the goldfish.

Hydromechanical models of steady swimming
Taylor's resistive model

Taylor (1952) developed a quasi-state resistive model of swimming. He assumed that: (1) the body of the animal can be likened to a long, thin cylinder of uniform cross-section; (2) the animal is propelled by propagating uniform, large-amplitude waves of constant wavelength and speed down its body; (3) each segment of the body executes a simple harmonic motion; (4) the forces acting on a segment at any given instant are given by the steady flow forces; and (5) flow is laminar.

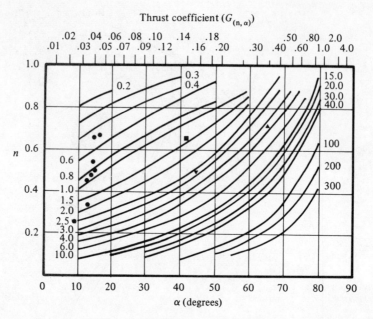

Fig. 76   Isopleths of $G_{(n,\alpha)}$ for various values of n and α. Drawn from Taylor (1952). Points are for *Salmo gairdneri* (●), *Gadus merlangus* (■), *Ammodytes langeolatus* (▼) and *Anguilla vulgaris* (▲). From Webb (1975*a*).

Experimental data on the forces acting on cylinders at moderate Reynolds Numbers (about 10^4) set at various angles of attack are combined with equations of motion for a segment and expressions are developed from which the normal and tangential forces and rate of working of a segment can be calculated. The rate of working of a segment can be written as

$$P = \frac{5.4}{2} \rho dU^3 \left(\frac{\mu}{\rho dU} \right)^{\frac{1}{2}} G_{(n,\alpha)} \tag{99}$$

where d is the diameter of the cylinder and $G_{(n,\alpha)}$ is a thrust coefficient, based upon the parameters n and α, where $n = U/V$ and $\alpha = A/\lambda_b$ (where A is the amplitude of the wave). Isopleths of $G_{(n,\alpha)}$ for various values of n and α are given in Fig. 76.

Webb (1975*a*) has employed eqn 99 to calculate the swimming power of three anguilliform swimmers. Calculated values for the total power output of specimens of *Anguilla vulgaris* (1.96×10^{-5} W, $l = 0.07$ m, $U = 0.04$ m s^{-1}), *Centronotus gunnellus* (2.93×10^{-4} W, $l = 0.12$ m, $U = 0.11$ m s^{-1}) and *Ammodytes lanceolatus* (4.03×10^{-4} W, $l = 0.18$, $U = 0.07$ m s^{-1}) were compared to the theoretical frictional drag power calculated from the standard hydrodynamic equations of resistance for a flat plate of equivalent area. The calculated frictional drag powers (1.6×10^{-6} W, 5.3×10^{-5} W and 2.9×10^{-5} W for *A. vulgaris*, *C. gunnellus* and *A. lanceolatus*, respectively) were much lower than the total power output values calculated from eqn 34, indicating that the drag of the swimming fish is probably much greater than that predicted by the theoretical equations.

The discrepancy between the theoretical frictional drag power figures and the values calculated from Taylor's model may be larger than indicated above if high pressure forces occur due to separation along the edges of the long median fins. Strictly speaking, eqn 99 should only be applied to round-bodied animals; most fish are laterally compressed. Taylor's model departs from the swimming of real fish in at least two other important respects. Firstly, the amplitude (A) and transverse velocity (W) are not constant along the entire length of the body. In anguilliform swimmers the amplitude, and therefore the transverse velocity, increase up to a point about half-way along the length of the body. Secondly, interactions between segments are not considered. It is reasonable to suppose that the movements of any given segment will influence the flow over those next to it.

Slender-body theory

The reactive forces generated by the aeroelastic 'fluttering' of aircraft wings have traditionally been analysed by employing slender-body theory. Recently this theory has been developed for application to the swimming motions of fish (e.g. Lighthill, 1960, 1970, 1971; Wu, 1971*a*, *b*, *c*, *d*; Newman & Wu, 1973, 1974). The theory emphasizes the reactive forces that are generated between the fish and a small mass of water that comes into contact with its surface (virtual mass). These forces become important for laterally compressed sections and are neglected by the older resistive theories of swimming. Most adult fish are laterally compressed and swim at high Reynolds Numbers ($> 10^3$) where inertial effects dominate over resistive forces.

Lighthill (1960, 1969, 1970) developed a theory applicable to increasing small amplitude motions of fish swimming in the anguilliform and subcarangiform modes. In a further study (Lighthill, 1971) he extended the analysis to include large amplitude motions. This model (termed elongated-body theory by Lighthill) is more appropriate to actual fish motions and is outlined below.

The fish is set in a co-ordinate system x, z (see Fig. 77). Any point on the centreline of the fish (i) can be defined relative to the length of the fish. The wake is separated from the fish by a vertical plane Π, perpendicular to the caudal fin. This defines a half-space C which contains the fish (Fig. 77). It is assumed that the length of the fish remains constant and that the fish has a uniform cross-section. The tangential velocity component u is given by

$$u = \frac{\partial x}{\partial t}\frac{\partial x}{\partial i} + \frac{\partial z}{\partial t}\frac{\partial z}{\partial i} \tag{100}$$

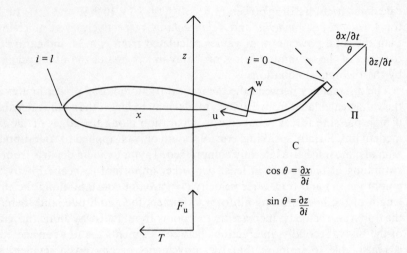

Fig. 77   Elongated-body theory. This figure is explained in the text. Based on Wardle & Reid (1977).

The normal velocity component w is

$$w = \frac{\partial z}{\partial t}\frac{\partial x}{\partial i} - \frac{\partial x}{\partial t}\frac{\partial z}{\partial i} \tag{101}$$

The w component has a large virtual mass (M) associated with it, the u component does not. The virtual mass is given by

$$M = \tfrac{1}{4}\rho\pi d^2\beta \tag{102}$$

where d is the cross-sectional depth of the section and β is a constant that depends on the shape of the section. For elliptic sections of any eccentricity $\beta = 1.0$. For most fish cross-sections β is close to 1. Values of β for a variety of cross-sectional shapes are shown in Fig. 78.

Considering firstly the momentum associated with the w motions, we can write the momentum per unit length of fish as

$$Mw\left(-\frac{\partial z}{\partial i}, \frac{\partial x}{\partial i}\right) \tag{103}$$

The momentum in C can be written as the integral of eqn 103,

$$\frac{\mathrm{d}}{\mathrm{d}t}\int_0^l Mw\left(-\frac{\partial z}{\partial i}, \frac{\partial x}{\partial i}\right)\mathrm{d}i \tag{104}$$

The rate of change of this integral can be written in terms of three components: 1. rate of change due to convection of momentum out of C across Π; 2. rate of change due to pressure forces acting across Π; and 3. the reactive force (T, F_s) acting on the fish. The third component must be subtracted from the sum of the other two, so

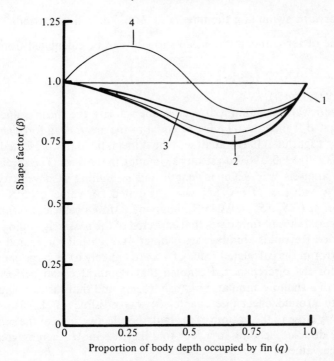

Fig. 78 Virtual mass coefficient, β, for a circular cross-section (1), a section with equal median fins (2), a fin-section with fin depths in the ratio $3:1$ (3) and for one fin only (4) is plotted against the proportion of the body depth that is occupied by fin (q). Based on Lighthill (1970).

$$\frac{d}{dt}\int_0^l Mw\left(-\frac{\partial z}{\partial i}, \frac{\partial x}{\partial i}\right)di = \left(-uMw\left(-\frac{\partial z}{\partial i}, \frac{\partial x}{\partial i}\right)+\tfrac{1}{2}Mw^2\left(\frac{\partial x}{\partial i}, \frac{\partial z}{\partial i}\right)\right)_{i=0} -(T, F_s)$$

(105)

Eqn 105 can be rewritten as

$$(T, F_n) = \left(Mw\left(\frac{\partial z}{\partial t}, \frac{\partial x}{\partial t}\right)-\tfrac{1}{2}Mw^2\left(\frac{\partial x}{\partial i}, \frac{\partial z}{\partial i}\right)\right)_{i=0} - \frac{d}{dt}\int_0^l Mw\left(-\frac{\partial z}{\partial i}, \frac{\partial x}{\partial i}\right)di \quad (106)$$

Essentially eqn 106 says that the force exerted by the fish on the water (T, F_n) is equal to the total rate of change of momentum within C minus the rate of change of momentum caused by the boundary Π.

The thrust (T) and tangential force (F_n) are given by

$$T = \left(Mw\frac{\partial z}{\partial t}-\tfrac{1}{2}Mw^2\frac{\partial x}{\partial i}\right)_{i=0} + \frac{d}{dt}\int_0^l Mw\frac{\partial z}{\partial i}di$$

(107)

$$F_n = \left(-Mw\frac{\partial x}{\partial t}-\tfrac{1}{2}Mw^2\frac{\partial z}{\partial i}\right)_{i=0} - \frac{d}{dt}\int_0^l Mw\frac{\partial x}{\partial i}di$$

(108)

In steady forward swimming the integral $\int_0^l Mw\dfrac{\partial z}{\partial i}\,di$ will be periodic with a mean value of zero, and so the mean thrust $\bar{T}$ can be calculated from

$$\bar{T} = \overline{\left(Mw\frac{\partial z}{\partial t} - \tfrac{1}{2}Mw^2\frac{\partial x}{\partial i} \right)_{i=0}} \tag{109}$$

(Lighthill, 1971, eqn 7).

In steady forward swimming at a constant velocity the mean value for thrust calculated from eqn 109 must be equal to the mean drag force acting on the fish. Lighthill (1971) analysed the kinematic data presented by Bainbridge (1958, 1960, 1963) on steady swimming in the dace. The specimen selected for analysis was 0.3 m in length and swimming at a velocity of 0.48 m s^{-1}. A value of $\bar{T} = 0.2$ N was calculated. A value of 5 N was calculated for $\tfrac{1}{2}\rho U^2 S_w$ ($S_w = 0.04$ m^2), implying a drag coefficient of about 0.04. This figure is about four times that expected of the passively gliding fish at the observed Reynolds Number (see chapter 4). Lighthill considered that sources of error in the calculated value of $\bar{T}$ were unlikely to be large enough to account for the discrepancy. He noted that Bainbridge had performed experiments in a shallow annular tank (fish wheel), and that thrust augmentation due to ground effect (see chapter 8) was possible. If the fish was swimming very close to the bottom of the tank the effective span of the caudal fin would be doubled, leading to a four-fold increase in M, thereby greatly increasing the value of $\bar{T}$.

If the discrepancy between inferred C_D values and those calculated from dead-drag measurements is real (i.e. not due to errors in experimental design or in analysis), the viscous drag of a swimming fish performing undulatory motions of its body must be higher than that of the equivalent rigid body. Bone (in Lighthill, 1971) suggests that drag augmentation could be caused by boundary layer thinning due to the oscillation of the body.

When a flat plate moves steadily in its own plane the boundary layer, if laminar, has a thickness of about $3\sqrt{(vx/U)}$ at a distance x from the leading edge. However, when the plate is moved perpendicular to itself the transverse motion results in a much thinner boundary layer (of the order of $0.6\sqrt{(vs/U)}$, where s is the span of the plate). The values of $\sqrt{(vx/U)}$ and $\sqrt{(vs/U)}$ would not be very different for a given value of U, and so the two boundary layer thickness values should differ by a factor of about five. A reduction in boundary layer thickness of a factor of five should result in an increase in the frictional drag of about the same magnitude.

Wardle & Reid (1977) employed eqn 109 to calculate the power output of a cod (*Gadus morhua*) over a range of swimming speeds. The fish had a mass of 3.22 kg and was 0.73 m in length. At the maximum swimming speed recorded (3.0 m s^{-1}) a maximum power output value of 90 W (or 30 W kg^{-1}) was calculated. Values ranged from about 10 W kg^{-1} to 30 W kg^{-1} over a speed range of about 2.0–3.0 m s^{-1}. In Table 1, drag coefficients implied by

Table 1 *Drag augmentation factor* $(C_{f(inf)}/C_{f(turb)})$ *for a 0.73 m cod. All notation is defined in the text. Based on data from Wardle & Reid* (1977)

U (m s^{-1})	$\bar{T}$ (N)	R_1 ($\times 10^6$)	D_f	$C_{f(inf)}$	$C_{f(turb)}$	$\dfrac{C_{f(inf)}}{C_{f(turb)}}$
2.56	5.34	1.869	698.48	0.0076	0.0040	1.90
2.47	4.34	1.803	650.23	0.0067	0.0040	1.68
2.48	5.11	1.810	655.51	0.0078	0.0040	1.95
2.46	3.32	1.796	644.98	0.0051	0.0040	1.28
2.54	5.53	1.854	687.61	0.0080	0.0040	2.00
2.55	5.39	1.862	693.04	0.0078	0.0040	1.95
2.53	6.32	1.847	682.21	0.0093	0.0040	2.33
2.54	4.04	1.854	687.61	0.0059	0.0040	1.48
2.70	8.13	1.971	776.97	0.0105	0.0040	2.63
2.71	9.82	1.978	782.73	0.0125	0.0040	3.13
2.71	16.89	1.978	782.73	0.0216	0.0040	5.40
2.71	16.79	1.978	782.73	0.0215	0.0040	5.38
2.91	21.96	2.124	902.53	0.0243	0.0039	6.23
2.96	17.74	2.161	933.81	0.0190	0.0039	4.87
2.74	10.24	2.000	800.16	0.0128	0.0040	3.20
2.67	16.76	1.949	759.80	0.0221	0.0040	5.53
2.67	17.10	1.949	759.80	0.0225	0.0040	5.63
2.96	15.87	2.161	933.81	0.0170	0.0039	4.36
1.88	6.43	1.372	376.70	0.0171	0.0043	3.98
1.93	8.56	1.409	397.00	0.0216	0.0042	5.14
1.91	8.29	1.394	388.81	0.0213	0.0043	4.95
1.97	13.21	1.438	413.63	0.0319	0.0042	7.60
2.01	7.98	1.467	430.59	0.0185	0.0042	4.41
2.10	7.28	1.533	470.02	0.0155	0.0042	3.69
2.07	5.88	1.511	456.68	0.0129	0.0042	3.07
2.85	19.63	2.081	865.70	0.0027	0.0039	5.82
2.88	31.16	2.102	884.02	0.0352	0.0039	9.03
9.93	17.32	2.139	914.98	0.0189	0.0039	4.85
3.02	21.74	2.205	972.05	0.0224	0.0039	5.74
3.97	9.78	2.168	940.13	0.0104	0.0039	2.67

Wardle & Reid's thrust force figures are compared with the theoretical frictional drag coefficients for a turbulent boundary layer $(C_{f(inf)}/C_{f(turb)})$. Discrepancies between the implied drag coefficients for the swimming fish and the rigid-body frictional drag coefficients of the same order as those found by Lighthill for dace are calculated.

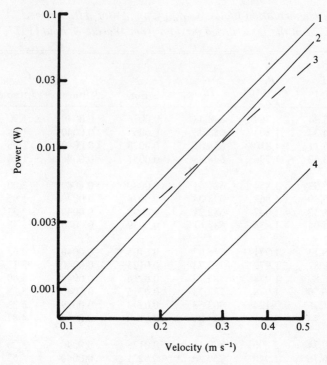

Fig. 79   Graph of power against speed for *Salmo gairdneri* ($l = 0.28$ m). Mean total power (1), mean useful power (2), useful power calculated from oxygen consumption measurements (3) and expected frictional drag power (4) are shown. Based on Alexander (1977).

Alexander (1967) analysed oxygen consumption data on *Onchorhynchus*, *Lepomis*, and *Carassius* (based on the results of Brett, 1964, Brett & Sutherland, 1965, and Smit, 1965, respectively). Alexander calculated that the power required to overcome friction drag in these fish is only about 2.5–6.0% of the total power released by metabolism. Four possible reasons for these low figures were considered: (1) a certain proportion of the total power released by metabolism is required for basic maintenance; (2) the muscles are only about 10–20% efficient at converting chemical energy into mechanical work; (3) the mechanical power exerted by the fish may exceed the values predicted on the basis of rigid-body drag; and (4) the total drag on the fish includes pressure drag as well as friction drag. Factors 1 and 4 will only account for a small proportion of the overall energy requirement of the fish. Assuming a muscle efficiency of 20%, Alexander calculated that the actual swimming drag of the fish would be between three and five times the estimated friction drag values.

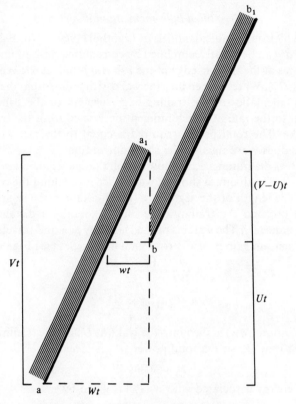

Fig. 80   Effect of trailing-edge segment on a water slice. This figure is explained in the text. Adapted from Lighthill (1969) and Webb (1975*a*).

Alexander (1977) employed a simplified version of Lighthill's model to analyse Webb's data on steady swimming in the trout (Webb, 1971*a, b*). The results of Alexander's calculations are shown in Fig. 79. The mean total mechanical power calculated from oxygen consumption data and the expected power requirement (i.e. that required to overcome friction drag) are plotted against swimming speed. There is good agreement between the mean useful power curve based on the hydromechanical analysis and that based on the metabolic data. However, the curve for mean useful power is much higher than the expected power curve.

Simplified bulk-momentum model

The detailed hydromechanical models of Lighthill (1960, 1970, 1971) and Wu (1971a, b, c, d) can be greatly simplified by considering bulk-momentum and energy changes at the trailing edge of the fish (Lighthill, 1969; Webb, 1975a; Alexander, 1977). Water is shed into the wake with momentum that is defined only by the kinematics of the trailing edge segment of the fish. Thrust is proportional to the rate at which momentum is shed from the trailing edge into the wake. The total power required is equal to the rate at which the trailing edge does work against the shed momentum.

Fig. 80 shows diagrammatically how a trailing edge segment affects a water slice close to it. The diagram shows a water slice just behind the trailing edge at a time t. The length of the segment is Vt. In a time t, the segment travels from aa_1 to bb_1 and the trailing edge moves laterally a distance Wt and forward a distance Ut. The water is pushed laterally at w and travels a distance wt. The momentum gain is Mw. From similar triangles (see Fig. 80) we have

$$\frac{wt}{Wt} = \frac{(V-U)t}{Vt} \tag{110}$$

$$w = W\frac{(V-U)}{V} \tag{111}$$

The momentum (Mw) is shed into the wake at U and the trailing edge does work against it at W, so the total power, P_T, is given by

$$P_T = MwUW \tag{112}$$

The kinetic energy associated with w (P_K) is given by

$$P_K = \tfrac{1}{2}Mw^2U \tag{113}$$

and therefore the thrust power P is

$$P = P_T - P_K = M(wWU - \tfrac{1}{2}w^2U) \tag{114}$$

Webb (1975a, 1978) has applied the simplified bulk-momentum model to a variety of fish that swim in the anguilliform and subcarangiform modes. Webb points out that, strictly speaking, the simplified model outlined above is only applicable to small amplitude motions. However, Lighthill (1971) has shown that large amplitude motions increase the rate of energy loss to the wake by a factor of $1/\cos\theta$, where θ is the angle between the trailing edge and the axis of progression. For subcarangiform swimmers θ is about 60° and mean values for $\cos\theta$ over the tail-beat cycle are about 0.85 and $1/\cos\theta$ is equal to about 1.18.

Efficiency

Efficiency is a useful criterion by which the swimming performance of different fish may be compared. To fully evaluate the propulsive system, three types of efficiency have to be considered: (1) the mechanical efficiency of the

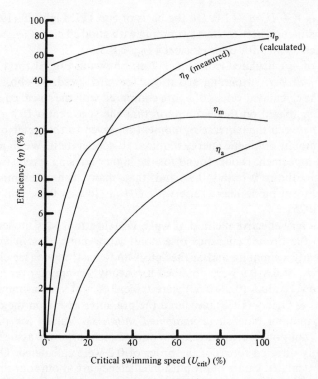

Fig. 81   Relations between overall aerobic efficiency (η_a), muscle efficiency (η_m) and propulsive efficiency (η_p) as a function of swimming speed (expressed as a percentage of the critical swimming speed) for *Salmo gairdneri*. From Webb (1975*a*).

caudal fin propeller, η_p; (2) the mechanical efficiency of the locomotor musculature, η_m; and (3) the overall aerobic efficiency, η_a.

The propulsive efficiency of a fish's flexural movements may be expressed as a Froude efficiency; we have

$$\eta_p = \frac{\text{Useful power used in overcoming drag}}{\text{Total rate of working of the body}} = \frac{P_T - P_K}{P_T} = \frac{P}{P_T} \quad (115)$$

or

$$\eta_p = 1 - \frac{\bar{P}_T - U\bar{T}}{\bar{P}_T} = \frac{1 - \frac{1}{2}\overline{(w^2)}_{i=l}}{\overline{(wW)}_{i=l}} \quad (116)$$

(Lighthill, 1970, eqn 12).

Provided that w and W at $i = l$ are related by eqn 111, we can write

$$\eta_p = 1 - \frac{1}{2}\frac{V - U}{V} \quad (117)$$

So, for good efficiency w must be considerably smaller than W and V not too much greater than U. Eqn 117 implies that η_p will always be greater than 0.5

and that as $V \to U$, $\eta_p \to 1.0$. On the basis of eqn 117, Lighthill (1969, 1970) draws a distinction between fish swimming with good efficiency ($\eta_p > 0.5$) and those which swim with poor efficiency ($\eta_p < 0.5$).

Values of η_p calculated from eqn 117 are shown in Fig. 81 for a specimen of *Salmo gairdneri* swimming at various forward speeds (Webb, 1971*a*, *b*; 1975*a*). The calculated values of η_p are compared with the so-called measured values of η_p, based on the assumption that the speed ratio U/V represents the ratio between the increase in momentum given to the water by the fish and the amount of kinetic energy required to accelerate it. Webb points out that good agreement is only found for the higher cruising speeds (when U/V is between about 0.7 and 0.9) and that many fish swimming in the subcarangiform mode may cruise with U/V as low as 0.23, implying that $\eta_p = 0.2$–0.25.

Using a very effective method of wake visualization, McCutcheon (1977) calculated the Froude efficiency of a small zebra danio (*Brachydanio rerio*) during steady swimming and in the 'push-and-coast' mode (see chapter 8). Values of η_p of 0.8–0.9 were obtained for steady swimming over a range of speeds. Lower values (0.55–0.60) were typical of unsteady swimming.

Du Bois & Ogilvy (1978) measured the pressures acting on the caudal fin of free-swimming bluefish (*Pomatomus saltatrix*). Values for propulsion efficiency (tail power acting in the forward direction/power associated with the laterally directed force) of the order of 0.2 were calculated. Du Bois & Ogilvy's calculated values of propulsion efficiency are about four times lower than those calculated on the basis of elongated body theory. The reason for this is not clear.

The mechanical efficiency of the locomotor muscles can be expressed as

$$\eta_m = \frac{\text{power developed by the muscles}}{\text{rate at which metabolic energy is made available}} \quad (118)$$

In chapter 2 we noted that vertebrate striated muscles are about 20% efficient at converting metabolic energy into mechanical work. Muscle efficiency is also plotted in Fig. 81.

The overall aerobic efficiency is

$$\eta_a = \frac{\text{useful power used in overcoming drag}}{\text{total metabolic power delivered}} = \eta_m \eta_p \quad (119)$$

Strictly speaking the work involved in support systems such as osmoregulation and ventilation should be subtracted from the denominator in eqn 119 for accurate estimates of η_a to be made. Values of η_a for the trout are plotted in Fig. 81. Values of η_a for subcarangiform swimmers are commonly within the range of 15 to 30% (Webb, 1975*a*).

Hydromechanical models of unsteady swimming

Fast-starts

Weihs (1973a) has extended Lighthill's large-amplitude elongated-body theory to cover the rapid acceleration of anguilliform and subcarangiform swimmers during fast-starts. The co-ordinate system and velocity relations are the same as those outlined in the section on slender-body theory. Eqn 107 (for the forwardly directed force acting on the fish) can be written as

$$T = - \frac{\mathrm{d}}{\mathrm{d}t} \int_0^l M w N \, \mathrm{d}i + (-uMwN - \tfrac{1}{2}Mw^2 S)_{i=0} \qquad (120)$$

where N and S represent normal and transverse co-ordinates of the centreline, respectively.

During fast-start behaviour vorticity is shed by sharp-edged surfaces such as fins and the forces associated with this must also be taken into account. We can write

$$T = - \frac{\mathrm{d}}{\mathrm{d}t} \int_0^l M w N \, \mathrm{d}i - \sum_{e=1}^{k} L_e \qquad (121)$$

where k is the number of sharp-edged surfaces and L_e is the momentum-shedding force associated with them. L_e is given by

$$L_e = \tfrac{1}{2}\rho S v^2 C_{\mathrm{L}(\alpha)} \alpha \qquad (122)$$

where S is the fin area, v is the velocity of the centre of pressure of the fin, $C_{\mathrm{L}(\alpha)}$ is the rate of change of the lift coefficient with the angle of attack α. The moment acting about the centre of mass of the fish is equal to

$$- \frac{\mathrm{d}}{\mathrm{d}t} \int_0^l M(\mathbf{r} w N) \, \mathrm{d}i - \sum_{e=1}^{k} (\mathbf{r} L_e + N_e) \qquad (123)$$

where $\mathbf{r}$ is a radius vector from the fish's centre of mass to the centre of pressure and N_e is the aerodynamic moment acting on the surfaces. Weihs applied the theory outlined above to calculate the forces and moments acting on a trout performing an L-type fast-start. Good agreement was found between accelerations predicted by the theory and experimental observations.

Turns

Weihs (1972) applied essentially the same analysis outlined in the previous section to the turning sequence of a rudd (see Fig. 75). He found that large normal forces were generated during stage two of the turn (see pp. 92–3 for a description of turning kinematics) and that a large anti-clockwise couple was produced. The last stage of the turn (stage three) was characterized by large forces in the final direction of movement of the fish. Weihs obtained similar results for the turning behaviour of the goldfish. From the results of his analysis Weihs was able to calculate values for the predicted radius of curvature of the turn. The predicted value for the radius of curvature for the rudd during stage two of its turn was 0.06 m, the observed value was 0.068 m.

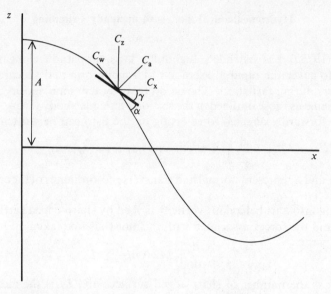

Fig. 82　Force coefficients in resistive model of whale swimming. This figure is explained in the text. From Parry (1949).

Hydromechanical models of thunniform swimming

Resistive model

Parry (1949) developed a resistive quasi-static model of lunate tail swimming based on a theory of bird flight originally due to von Holst & von Kuchemann (1942). Von Holst & von Kuchemann considered a wing of semi-span σ, moving at a velocity U whilst oscillating sinusoidally at 90° to U with frequency f and amplitude A. The angle of attack of the wing varies about a mean value α_0. To obtain the force C_x and C_z in the direction of movement and at right angles to it, von Holst & von Kuchemann first obtained expressions for the forces C_w and C_a (see Fig. 82)

$$C_a = \frac{2\pi\alpha}{1+(2/\text{A.R.})} \tag{124}$$

$$C_w = \frac{C_a^{\ 2}}{\pi\text{A.R.}} + \text{profile drag} \tag{125}$$

Parry (1949) showed that for the flukes of *Tursiops truncatus*, α is approximately 90° out of phase with the tail's oscillation and $\alpha_0 = 0$. So, the mean force in the x-direction is given by

$$\bar{C}_x = \frac{4}{\lambda}\int_0^{\lambda/4}(C_a\sin\gamma - C_w\cos\gamma)\,\mathrm{d}x \tag{126}$$

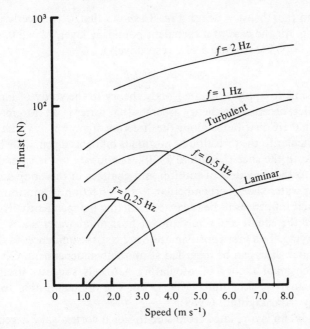

Fig. 83   The relation between thrust, drag and swimming speed in *Tursiops truncatus* ($l = 1.8$ m). From Parry (1949).

If γ is small, $\sin \gamma = 0$ and $\cos \gamma = 1$. Therefore

$$\bar{C}_x = C_a \alpha_1 \pi A \, (f\sigma/U) - \tfrac{1}{2} C_a^2 \alpha_1^2 \, (1/\pi A.R.) - C_{Do} \qquad (127)$$

where α_1 is the amplitude of the oscillatory change in the angle of attack, C_p is a profile drag coefficient and the term $(f\sigma/U)$ is the reduced frequency parameter. $\bar{C}_z$ is given by

$$\bar{C}_z = \frac{4}{\lambda} \int_0^A C_a \alpha \, dz = \frac{C_a \alpha_1 \pi}{4} \qquad (128)$$

We can write the forward thrust force T as

$$T = \tfrac{1}{2} \rho S_w U^2 \bar{C}_x \qquad (129)$$

Parry measured S_w, A.R., A and σ for a specimen of *T. truncatus* and found that $S_w = 0.034 l^2$, A.R. $= 2$, $A = 1.7$ and $\sigma = 0.31 l$. Values of $\alpha_1 = 10°$ and $C_p = 0.023$ were assumed. The thrust force was then expressed in terms of l, f and U

$$T = 0.0175 l^2 U^2 \frac{0.38 lf}{U} - 0.047 \qquad (130)$$

Thrust is plotted against U for various values of f in Fig. 83. Friction drag was calculated for the case of fully developed laminar and turbulent boundary layer types. The friction drag curves are also plotted on Fig. 83. Points at which the drag curves intersect the thrust curves indicate conditions of equilibrium (i.e. thrust = drag). Fig. 83 shows that U is directly proportional to f (e.g. for the case of a turbulent boundary layer, $U = 2.0$, 4.0 and 8.0 m s^{-1} for $f = 0.25$, 0.5 and 1 Hz, respectively).

Reactive model

It is not possible to apply elongated-body theory to the study of lunate tail hydrodynamics, because the theory assumes that water is set into motion by movements that are distributed along the direction of motion. A vertical water slice perpendicular to the direction of motion is influenced primarily by body motions close to the slice (Lighthill, 1970). The lunate tail is spread out at right angles to the direction of motion and changes in the body action on neighbouring water slices perpendicular to the direction of motion are too abrupt for their influences to be regarded as acting independently. However, the action of the lunate tail's motion on horizontal water slices is more gradually varying. To a first approximation at least, the influences of different horizontal water slices can be regarded as mutually independent. Given this, the two-dimensional theory of oscillating aerofoils suggests itself as an approximate analysis (Lighthill, 1969, 1970). The discussion that follows is drawn largely from Lighthill (1969).

As the caudal fin is oscillated from side to side a vortex wake is generated. The wake is made up of alternating clockwise and anti-clockwise vortex rings. In order to produce thrust a wake must be generated. If thrust is to be produced efficiently the energy loss into the wake must be minimized. Thrust increases as the first power of vortex strength, however energy losses vary as its square. Calculation of thrust or energy loss to the wake takes account of how water movements induced by wake vorticity modify the fin's effective angle of attack. This is done by calculating the pressure distribution. The gradient of the pressure distribution at right angles to the fin is equal to the rate at which the fin's movement is causing local water momentum in that direction to change (Wu, 1961).

As in elongated-body theory the energy lost in the vortex wake can be minimized by partially feathering the fin. The feathering parameter θ is a number, usually less than one. When $\theta = 1$ the fin is geometrically feathered. The fin oscillates at a velocity W and with a maximum angle of attack $\theta W/U$ that is reached in the same phase. The thrust generated per unit area of the fin can be divided by $\frac{1}{2}\rho W^2$ to obtain a thrust coefficient C_T. Efficiency η simply expresses the proportion of the power exerted by the fish that goes into providing thrust. An example of the results obtained by employing the two-dimensional theory is shown in Fig. 84. Fig. 84 illustrates the inverse relation between C_T and η as θ increases.

Lighthill (1969, 1970) points out that the two-dimensional theory will tend to overestimate efficiency because it only takes account of the energy of the cross-section of the wake vorticity (at right angles to the direction of motion)

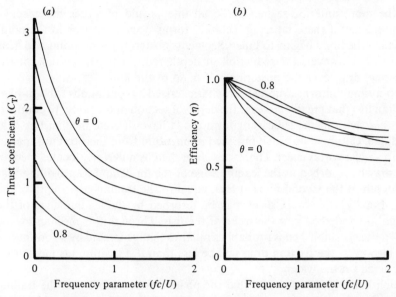

Fig. 84   (*a*) Relation between the thrust coefficient C_T and the reduced frequency parameter (fc/U) showing the influence of the feathering parameter θ. (*b*) Efficiency is plotted against the frequency parameter for various values of θ. From Lighthill (1969).

and does not take account of trailing vorticity parallel to the direction of motion. Lighthill's theory has been generalized by Chopra (1974*a*, *b*) who discusses the influence of varying chord, aspect ratio, feathering and position of the pitching axis on thrust and energy losses.

Kinematics, hydrodynamics and functional design

The magnitude of the propulsive thrust force produced by a fish in steady forward swimming is directly influenced by a range of morphological features that affect both kinematics and cross-sectional properties such as virtual mass. Lighthill (1969, 1970) noted several morphological types among anguilliform swimmers.

In some anguilliform swimmers the posterior third or so of the body tapers towards the trailing edge (e.g. *Lophotes, Gastrostomus*). The reduction in body depth is associated with a decrease in virtual mass, and only a slight increase in transverse velocity. Thrust is proportional to d^2 (see eqn 102) and only to the first power of W. The more anterior segments have relatively large values of d, values of W closely approach the maximum value that is reached at the trailing edge. This implies that most of the useful work of swimming is done by the more anterior segments. To all intents and purposes the effective trailing edge of these tapering 'eel-like' forms occurs anterior to the point at which the body begins to taper. Segments posterior to this point add little to thrust. However, the reduction in depth reduces the magnitude of the frictional drag over the posterior half or so of the body.

Many anguilliform swimmers are characterized by continuous 'ribbon-like' median fins that greatly increase the depth of section of the body. In contrast, fish that swim in the subcarangiform mode have discrete median fins (e.g. Gadidae, Carangidae). Such fins shed momentum from their trailing edge in the form of a vortex sheet. The vortex sheet shed by anteriorly located median fins may be resorbed at the leading edge of the fin behind it, providing that the depth of the second fin is at least equal to the trailing-edge depth of the first. Finally, the vortex sheet may be resorbed by the leading edge of the caudal fin and shed into the wake of the fish. Through this mechanism the flow pattern established is probably similar to that generated by a continuous fin. However, the friction drag will be less than if a continuous median fin had acted on the water.

Lighthill (1970) has shown that the phase difference between the trailing edge of an anterior fin and the leading edge of the next posterior fin must be less than $0.5\lambda_B$ for effective resorption of the vortex sheet to occur. The phase difference is given by

$$\frac{2\pi(x_2 - x_1)}{\lambda_B} \tag{131}$$

where x_2 is the position of the leading edge of the posterior fin and x_1 is the position of the trailing edge of the anterior fin. However, the vortex sheet moves downstream at U and the propulsive wave travels at the greater velocity V and so the actual phase difference between the motions of the body and the vortex sheet is given by

$$\frac{2\pi(x_2 - x_1)}{\lambda_B}\left(\frac{U}{V} - 1\right) \tag{132}$$

Webb (1975a, 1978) points out that effective vortex-sheet resorption between the median and caudal fins probably occurs in sharks. Webb analysed a film record made by Gray (1933a) of *Acanthias vulgaris* swimming at 0.178 m s^{-1} with $V = 0.256$ m s^{-1}. The fish was 0.46 m in length and λ_B was measured at 0.30 m. The trailing edge of the first dorsal fin was about 0.194 m from the nose of the fish. The first dorsal fin was followed by a smaller, second dorsal fin. However, it is likely that effective vortex-sheet resorption does not occur until the leading edge of the caudal fin. The caudal fin leading edge was measured at a distance of 0.371 m from the nose. This gives a value for θ of 0.52π. Similar values can be calculated for the phase difference of the median fins and caudal fin of the sturgeon (*Acipenser sturio*) from Hertel's data on sturgeon swimming kinematics (Hertel, 1966).

Lighthill (1969, 1970) explains the swept-back planform of the caudal fin of subcarangiform swimmers by proposing that the 'scooped out' region of the fin is filled by a vortex sheet. The caudal fin should function as if it were structurally complete but be subject to less frictional drag than a non-swept-back fin.

Among fish that swim in the subcarangiform and carangiform modes there is a general trend towards lateral compression and a concentration of thrust forces towards the caudal fin which undergoes large amplitude oscillations and produces most of the thrust. The large amplitude motions generate substantial side-forces. Lateral forces are almost balanced for fish that swim in the anguilliform mode, where a complete wavelength may be present on the body. Yawing of the anterior end of the body due to unbalanced recoil forces could greatly increase the energy cost of swimming in subcarangiform and carangiform fish.

Lighthill (1969, 1970) argues that two morphological characteristics of carangiform swimmers tend to reduce the influence of recoil forces; they are: 1. increased depth of section over the centre of mass; and 2. reduced depth of section posteriorly (narrow necking).

The recoil forces generated by the caudal fin are resisted by a large moment of inertia due to the large virtual mass effect of the anterior portion of the body. This effect may be enhanced due to the presence of median fins over the centre of mass. Body depth decreases progressively from a point over the centre of mass to the caudal peduncle in many fish that swim in the carangiform modes. Reduced local cross-sectional depth minimizes the magnitude of virtual mass effects posteriorly. In many Carangidae and Scombriodae the caudal peduncle is streamlined in its plane of oscillation and this further reduces recoil forces. Lighthill (1977) extends his large-amplitude elongated-body theory to the analysis of recoil effects in carangiform swimmers. Lighthill concludes that lateral oscillations of the anterior portion of the body should not exceed about 20% of the maximum cross-sectional depth for efficient swimming. Measurements from the film records of Gray (1933a) and Bainbridge (1963) indicate that the amplitude of the oscillations of the anterior portion of the body during steady forward swimming in subcarangiform fish does not exceed the critical value.

Breder (1926) and Gray (1968) noted that caudal fin amputation in carangiform swimmers does not markedly reduce steady swimming performance. The influence of caudal fin amputation on steady swimming performance was investigated further by Webb (1973a). Webb examined the effects of partial caudal fin amputation on the kinematics and metabolic rate of the sockeye salmon (*Oncorhynchus nerka*). Tail-beat frequency, trailing edge amplitude, swimming speed and oxygen consumption rate were all unaffected by removal of either the upper or lower lobes of the caudal fin. However, significant reductions in the 45-min U_{crit} were found when both lobes of the caudal fin were amputated.

The optimal morphological requirements for good unsteady swimming performance are: 1. large caudal fin and body area; 2. deep caudal peduncle; 3. large depth of section over the centre of mass; 4. flexible body; and 5. large ratio of muscle mass to overall body mass.

Hydromechanical theory predicts that large fins are required for good acceleration and turning performance (Weihs, 1972, 1973*a*). To test this, Webb (1977) studied the effects of median fin amputation on the fast-start performance of trout (*Salmo gairdneri*). A trend towards decreasing performance was found for the following sequence of amputations: control (pelvic fins amputated), dorsal fin, anal fin, dorsal lobe of caudal fin and ventral lobe of caudal fin, ventral lobe of caudal fin and anal fin. The results of Webb's experiments confirms the predictions of the theory concerning the importance of large lateral body profile for good fast-start performance.

Large amplitude body movements, correlated with a large depth of section, can be expected to generate very large recoil forces. It has been noted that the oscillation of the anterior end of the body caused by such forces can be energetically wasteful (Lighthill, 1969, 1970, 1977). Large anterior body depth and mass is required to reduce the effect of recoil forces during unsteady swimming. The pike (*Esox*) lacks any anterior depth or mass enhancement and is subject to large recoil movements of its centre of mass during fast-starts (Webb, 1978, 1981). In contrast the sunfish (*Lepomis*) shows good fast-start performance and is not subject to large recoil movements (Webb, 1978).

At first sight it would seem reasonable to suppose that a large mass of propulsive musculature would be required for good fast-start performance. However, good laterally compressed profiles are associated with a relatively low muscle mass, good profile characteristics are not compatible with a large muscle mass. To all intents and purposes, the advantages of good longitudinal depth distribution and a large proportion of propulsive musculature tend to cancel each other out. Forms with good lateral profile characteristics (e.g. *Cottus*, *Etheostoma*) do not show better fast-start performance than more round-bodied forms such as the trout.

From the foregoing discussion it is apparent that most of the morphological requirements for good steady and unsteady swimming performance are mutually exclusive. Fast-start performance is extremely important to most fish; it determines the success of predator strikes and the success of prey escape attempts. Webb (1977, 1978, 1982) points out that most non-scombriod fish are adapted for good fast-start performance.

6

Mechanics of 'non-body' modes of swimming

Introduction

Most of our discussion of fish locomotion to this point has been concerned with fish that swim in the anguilliform and carangiform modes. Many fish are not propelled by undulations of the body and caudal fin however, but employ the action of their median and/or paired fins for most of their swimming behaviour. Such forms are said to swim in the 'non-body' modes. Among the 'non-body' swimmers, Breder (1926) proposed the term labriform (after the family Labridae) to describe all forms of oscillatory pectoral fin swimming. In addition, six modes of undulatory fin propulsion were described, based on the morphological position of the active fin or fins.

In amiiform swimming (after the genus *Amia*) the fish is propelled by the dorsal fin (e.g. *Amia, Hippocampus, Gymnarchus*). Forms that are propelled by the anal fin (e.g. *Gymnotus, Hyperopsus, Notopterus*) are assigned to the gymnotiform mode. Balistiform propulsion (after the triggerfishes of the family Balistidae) involves a combination of the anal and dorsal fins (e.g. *Balistes, Monocanthus*). In the diodontiform, tetraodontiform and rajiform modes the main propulsive force is produced by undulatory pectoral fins (e.g. *Diodon, Tetradon* and *Raja*, respectively).

Here, a basic distinction is made between forms that are propelled by undulatory fins and those that are propelled by oscillatory fins. Undulatory fin forms are divided into two groups on the basis of fin kinematics. The kinematics of oscillatory and undulatory propulsion are discussed and hydromechanical models are outlined that can be employed in the calculation of swimming thrust, power and efficiency. Undulatory fin swimming is discussed in terms of slender-body theory and actuator-disc theory. A distinction is made between oscillatory fin swimmers that are propelled on a drag-based (rowing) principle and those that are propelled by a lift-based mechanism. Both groups are analysed by blade-element theory.

Applications of the hydromechanical models are discussed with reference to the morphology, behaviour and ecology of the forms concerned. Finally, aspects such as speed, acceleration, manoeuvrability, streamlining and drag in 'non-body' swimmers are discussed in relation to other fish.

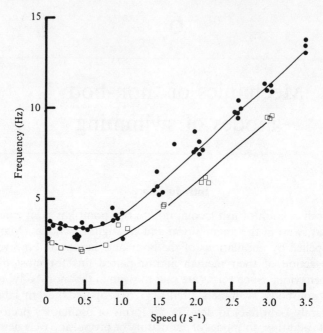

Fig. 85   Fin-beat frequency is plotted against specific swimming speed for the dorsal (●) and anal (□) fins of the triggerfish *Rhinecanthus aculeatus*. From Blake (1978).

Kinematics of undulatory fin propulsion

The classification of undulatory fin swimming modes proposed by Breder (1926) has gained a degree of general acceptance. This is probably due to the fact that it is based on a small number of readily determined characters, such as fin number and position. Breder's classification does not relate to any functional aspects of fin kinematics however, and consequently forms that differ fundamentally in this respect are arbitrarily assigned to the same grouping. For example, both the seahorse (*Hippocampus*) and electric-eel (*Gymnarchus*) are propelled by their undulatory dorsal fin, and therefore fall into the amiiform mode of the classification. However, it can be argued that the dorsal fin kinematics in these forms represent distinct end points in a continuum with respect to fin waveform frequency, amplitude and wavelength.

We will employ a simple binary classification of undulatory fin propulsion based on fin kinematics:

Group One: Forms with fins which show waveforms of relatively high amplitude, low frequency and large wavelength (e.g. Tetraodontidae, Diodontidae, Ostraciidae, Gymnarchidae).

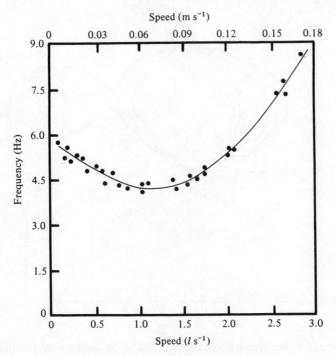

Fig. 86   Fin-beat frequency is plotted against swimming speed for the pectoral fins of *Cymatogaster aggregata*. From Webb (1974).

Group Two: Forms with fins of low amplitude, high frequency and small wavelength (e.g. Synathidae, Aluteridae, Canthigasteridae).

It is interesting to note that certain families contain some genera that fall into Group One of the classification and others that belong to Group Two (e.g. Balistidae).

Fin kinematics for representatives of both groups have been described by Blake (1983a). Certain kinematic trends are common to both. For example, waveform frequency increases linearly with increasing swimming speed after speeds of about 1.0–1.5 l s^{-1} (see Figs 85 and 86). Amplitude is highly variable at low speeds in most of the forms examined. Wavelength tends to remain more or less constant over a large speed range. In contrast to this, values for the three principal kinematic parameters (frequency, amplitude and wavelength) for Group One and Group Two forms are very different.

The dorsal fin kinematics of *Hippocampus* (Group Two) have been compared with those of *Gymnarchus* (Group One). Mean specific wavelength (mean wavelength of the waveforms present on the fin divided by the length of the fin base) was about 75% greater in *Gymnarchus* (0.51) than in *Hippocampus* (0.29). At similar swimming speeds (about 16 cm s^{-1}) waveform

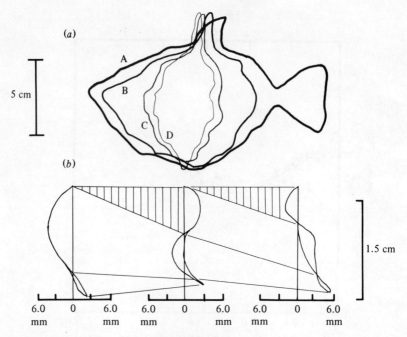

Fig. 87   Turning manoeuvre in *Rhinecanthus aculeatus*. Successive body positions are shown in diagram (*a*). The starting position is 1, 2 is after 0.04 s, 3 is after 0.11 s and 4 is after 0.17 s. The waveforms present on the fin during the turn are illustrated in (*b*). Successive diagrams are at intervals of 0.05 s. From Blake (1978).

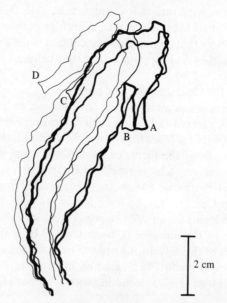

Fig. 88   A 180° turn to the left in the seahorse *Hippocampus hudsonius*. Position 1 is the starting position, 2 is after 0.06 s, 3 is after 0.105 s and 4 is after 0.285 s. From Blake (1976).

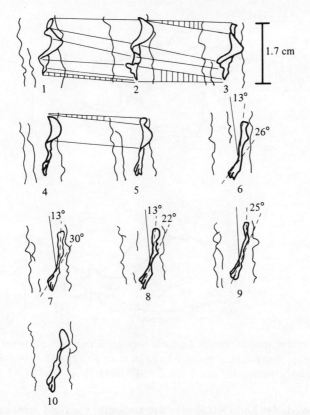

Fig. 89 This figure shows the changes in the long axis of the dorsal fin of *Hippocampus* during an 180° turn to the left. Frames 1–6 are at 0.015 s intervals. Diagram 7 is after 0.165 s, 8 is after 0.195 s, 9 is after 0.24 s and 10 is after 0.255 s. From Blake (1976).

frequency was about 16.4 times as great in *Hippocampus* than in *Gymnarchus* (41 Hz and 2.5 Hz, respectively).

Turning manoeuvres can be divided into two types on the basis of the way in which they are initiated:

Type One: Turns that are initiated by a deflection of the body to the left or to the right.

Type Two: Turns that are initiated due to a progressive deflection of one or more fins independently of the body which remains rigid.

Type One turns are performed in essentially the same way as those described in the previous chapter for fish that swim in the carangiform modes. Blake (1978) describes a 180° turn to the right in the triggerfish *Rhinecanthus aculeatus*. The turn and the waveforms present on the dorsal fin during the manoeuvre are reconstructed in Fig. 87.

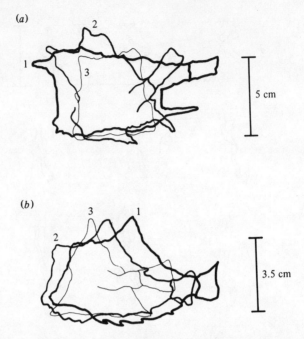

Fig. 90   Turning manoeuvres in *Lactoria cornuta* (*a*, time lapse between positions 1 and 2 is 0.35 s; the specimen took 0.82 s to go from 1 to 3) and *Tetrasomus gibbosus* (*b*, time lapse between positions 1 and 3 was 0.86 s). From Blake (1977).

The Synathidae and Ostraciidae are encased by a rigid armour and body flexure is therefore not possible. However, the median fins of these forms are able to change their orientation relative to the principal body axis; in addition, parts of the fins may move relative to others. Blake (1976) describes a 180° turn to the left in *Hippocampus* (see Fig. 88) in which the posterior of the dorsal fin was deflected progressively over to the right, until about two-thirds of the length of the fin was deflected (Fig. 89). Fig. 90 illustrates turning manoeuvres in the Ostraciidae, like the seahorse these forms are capable of rapid turning manoeuvres that involve little or no lateral translation of the body.

Kinematics of oscillatory fin propulsion

Pectoral fin rowing in the angelfish (*Pterophyllum eimekei*) is described by Blake (1979*a*, *b*, 1980*a*). In slow forward swimming the pectoral fins are the only active fins. In the power stroke the pectoral fins are oriented 'broadside on' to the flow at a high geometric angle of attack. The phase difference between the most dorsal and ventral fin rays is small and the fin appears to rotate about its base as a unit. Fin-beat frequency and amplitude both

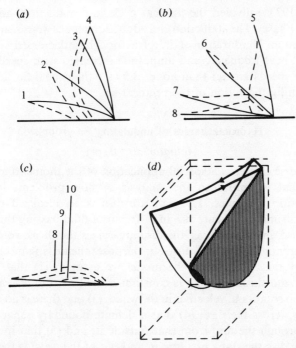

Fig. 91 Movements of the pectoral fin of *Cymatogaster aggregata* in pectoral fin pattern A (*a–c*, the interval between frames is 0.018 s) and a diagrammatic representation of the fin motion (*d*, arrows indicate the motion of the tip of the trailing edge of the fin which is indicated by shading). From Webb (1973*a*).

increase with increasing swimming speed. The product of frequency and amplitude is linearly related to speed. At the end of the power stroke the fins are 'feathered' and moved forward. In fast swimming the pectoral fins are held against the sides of the body and the fish is propelled by its caudal fin.

Lift-based pectoral fin propulsion has been described in the shiner seaperch (*Cymatogaster aggregata*) by Webb (1973*b*). The fins oscillate in an up-and-down motion and at the same time pass a wave back over their length. On the basis of an index relating the wavelength of the wave to the length of the trailing edge of the fin, Webb classified the fin movements into two groups, pectoral fin patterns A and B. In pattern A the wavelength of the waveform on the fins was about twice the trailing edge length, resulting in a phase difference of about π between the anterior and posterior fin rays. The phase difference in pattern B was only about 0.2π. Pattern A fin movements are illustrated in Fig. 91. Movements of the pattern A type were observed up to specific swimming speeds of about $2.0\ l\,\mathrm{s}^{-1}$, after which pattern B motions were recorded.

Webb (1973b) divided the fin-beat cycle into abduction, adduction and refractory phases. The abduction and adduction phases were of approximately equal duration. The duration of the refractory phase decreased with increasing speed. Fin-beat frequency and amplitude increased with speed. At specific swimming speeds greater than about $3.5 \, l \, s^{-1}$ the caudal fin became active, supplementing the action of the pectoral fins.

Hydromechanics of undulatory fin propulsion

Actuator-disc theory

Actuator-disc theory is a special application of the momentum principle, a concept that is central to the analysis of flow problems involving the determination of forces. The actuator-disc is an idealized device which produces a sudden pressure rise in a stream of fluid passing through it. The pressure rise, integrated over the disc area gives the thrust force associated with the driving mechanism. For our purposes the driving mechanism is the undulatory fin. In applying the model we assume: (1) that the pressure increment and thrust loading is constant over the entire disc area; (2) that there are no rotational velocities in the wake; (3) that there is no discontinuity in velocity across the disc; (4) that a definite boundary separates the flow passing through the disc from that outside it; and (5) that in front of and behind the disc the static pressure in and out of the wake is the same as the free-stream static pressure.

Far upstream of the fin the pressure is p_0 and the water velocity is U, the swimming speed. Just upstream of the fin the velocity is assumed to have increased to $(U+v)$ and in accordance with Bernoulii's theorem (see chapter 1) the pressure falls to p. Immediately downstream of the fin the pressure increases by Δp to $(p+\Delta p)$, while the velocity is unchanged. Far downstream of the fin the pressure returns to p_0, while the velocity is increased to (U/v^*). Applying Bernoulii's theorem to a region directly in front of the disc, we have

$$p+\tfrac{1}{2}\rho(U+v)^2 = p_0+\tfrac{1}{2}\rho U^2 \tag{133}$$

Similarly, for the region just behind the disc

$$p+\Delta p+\tfrac{1}{2}\rho(U+v)^2 = p_0+\tfrac{1}{2}\rho(U+v^*)^2 \tag{134}$$

Subtracting eqn 133 from 134 gives

$$\Delta p = \tfrac{1}{2}\rho(U+v^*)^2-\tfrac{1}{2}\rho U^2 = \rho v^*(U+\tfrac{1}{2}v^*)^2 \tag{135}$$

The thrust force is given by

$$T = \Delta p S_d = S_d \rho v^*(U+\tfrac{1}{2}v^*) \tag{136}$$

where S_d is the disc area (area swept-out by the fin rays). During hovering $U = 0$, and

$$T = \tfrac{1}{2}\rho S_d v^{*2} \tag{137}$$

The rate of mass flow through the disc is

$$S_d(U+v)\rho \tag{138}$$

and therefore the increase in momentum of the wake is given by

$$\rho S_d(U+v)v^* \tag{139}$$

Thrust equals the rate of change of momentum and

$$S_d\rho v^*(U+\tfrac{1}{2}v^*) = S_d\rho v^*(U+v) \tag{140}$$

so

$$v^* = 2v \quad \text{or} \quad v = \tfrac{1}{2}v^* \tag{141}$$

The minimum power required to generate the induced velocity v (induced power) P_{id} is

$$P_{id} = TU = S_d\rho v^*(U+\tfrac{1}{2}v^*) \tag{142}$$

For hovering P_{id} is given by

$$P_{id} = TU = \tfrac{1}{2}\rho S_d Uv^{*2} \tag{143}$$

However, the power input P_{in} is

$$P_{in} = T(U+v^*) \tag{144}$$

The ideal efficiency of the undulating-fin system can be written as

$$\eta = P_{id}/P_{in} = TU/T(U+v) = U/(U+v^*) \tag{145}$$

Flow velocities around the dorsal and pectoral fins of the seahorse (*Hippocampus*) and the mandarin fish (*Synchropus picturatus*) have been determined by particle-based flow visualization techniques (Blake, 1976, 1979c). Swimming power is plotted against swimming speed for a specimen of *Synchropus* in Fig. 92. We note that: 1. the induced power declines rapidly as forward speed increases; and 2. the power to overcome fin drag (profile power) and the drag on the body (parasite power) increase with speed. When the curves are summed, a U-shaped total power curve results. The mandarin fish is a negatively buoyant, demersal teleost and its power curves can be compared to those of other negatively buoyant animals (e.g. birds) and man-made heavier than air machines such as helicopters. The induced power curve is similar in form (see Pennycuick, 1968 and Rayner, 1979 on birds and Bramwell, 1976 on helicopters). In birds, parasite power increases rapidly with speed, to become the dominant power component at high speeds. Absolute swimming speeds are low in *Synchropus*, and parasite power does not become an important power component. The total power curve in helicopters (Bramwell, 1976) and some birds (Pennycuick, 1968; Tucker, 1973; Green-walt, 1975; Rayner, 1979) is U-shaped with a definite minimum power speed. Although basically U-shaped, the total power curve of *Synchropus* is rather flat. In this connection, it is interesting to note that *Synchropus* does not show a preferred swimming speed in captivity (Blake, 1979c).

Blake (1980b, 1983a) has developed a modified form of the theory that allows for the calculation of thrust, power and ideal efficiency in the absence

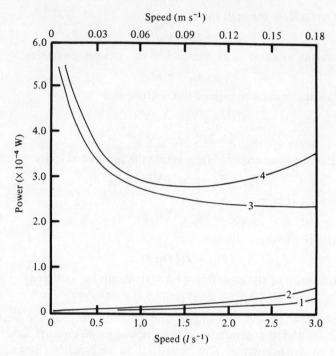

Fig. 92   Power in relation to swimming speed for *Synchropus picturatus* (notation: 1, parasite power; 2, profile power; 3, induced power; 4, total power). From Blake (1979c).

of detailed information on fin flow velocities. However, the drag force acting on the fish must be known. The model has been employed in an analysis of swimming mechanics in the seahorse and electric fish (*Gymnarchus*). Values of η for *Gymnarchus* of the order of 0.6–0.7 were common, those for *Hippocampus* rarely exceeded half of these values. The electric-eel falls into Group One of the kinematic classification, the seahorse is a Group Two form. Group One forms develop thrust more efficiently than Group Two forms, as it is more efficient when a large mass of water is accelerated to a relatively small eventual velocity than when a small mass of water is accelerated to relatively high velocities. *Gymnarchus* is an active carnivore (Sterba, 1962) and it is likely that the high ideal efficiency of its dorsal fin enables it to cruise economically in search of its prey. The seahorse, however, is a relatively sluggish fish that feeds on slow swimming zooplankton. It is probable that the overall energy budget of *Hippocampus* is not significantly affected by the relatively low ideal efficiency of its dorsal fin. It could be argued that the high frequency fin of the seahorse is adaptive, as it is likely that it operates at a frequency that is beyond the flicker fusion frequency of the eyes of its potential predators, thereby rendering it invisible to them.

Eqn 46 has been employed to calculate the induced power produced by the pectoral fins of *Synchropus* during hovering at various heights above the ground (Blake, 1979*d*). This work is discussed in chapter 8 where the 'ground effect' is considered.

Slender-body theory

We have already noted that slender-body theory has been applied successfully to the analysis of anguilliform and subcarangiform swimming (see chapter 5). Blake (1976) points out that the theory should also be applicable to the undulatory motions of median and paired fins. Details of slender-body theory are discussed in chapter 5, and only the modifications required for undulatory fin analysis will be stressed here.

Mean thrust can be calculated from an expression that is similar to that given in eqn 109; we can write

$$\bar{T} = \overline{[Mw(W - \tfrac{1}{2}w)]}_{i=R} \tag{146}$$

where R is the total fin length (i.e. $i = R$ refers to the trailing edge of a fin). Similarly, we can write expressions for the mean total power $\bar{P}_{\mathrm{T}}$, mean rate at which kinetic energy is lost to the wake $\bar{P}_{\mathrm{K}}$, and mean thrust power $\bar{P}$:

$$\bar{P}_{\mathrm{T}} = \overline{U[Mw(\partial h/\partial t)]}_{i=R} \tag{147}$$

$$\bar{P}_{\mathrm{K}} = \overline{U\left[\tfrac{1}{2}Mw\left(\frac{\partial h}{\partial t} + U\frac{\partial h}{\partial i}\right)^2\right]}_{i=R} \tag{148}$$

$$\bar{P} = \overline{U[Mw(\partial h/\partial t)]}_{i=R} - \overline{U\left[\tfrac{1}{2}Mw\left(\frac{\partial h}{\partial t} + U\frac{\partial h}{\partial i}\right)^2\right]}_{i=R} \tag{149}$$

The propulsive efficiency of the undulatory fin system can be determined from

$$\eta_{\mathrm{p}} = 1 - (\bar{P}_{\mathrm{T}} - U\bar{T})/\bar{P}_{\mathrm{T}} = 1 - \tfrac{1}{2}\overline{(w^2)}_{i=R}/\overline{(wW)}_{i=R} \tag{150}$$

Blake (1983*b*) has employed slender-body theory to calculate the swimming thrust, power and efficiency in various species of electric-eels (Gymnotidae) and knifefishes (Notopteridae). The value of the lateral velocity of the fin involved (anal fin) is zero at the base and increases to a maximum at the distal edge. Due to this, mean values of W and w were employed in the calculations. Two questions were considered:
1. What, if any, are the hydromechanical advantages of undulatory median-fin propulsion over anguilliform and subcarangiform swimming? 2. Is it possible to offer explanations based on the hydromechanical theory for the morphological similarities and differences between the various species considered?
The above questions were approached by considering the following criteria:
(1) propulsive efficiency; (2) speed and acceleration; (3) manoeuvrability; (4)

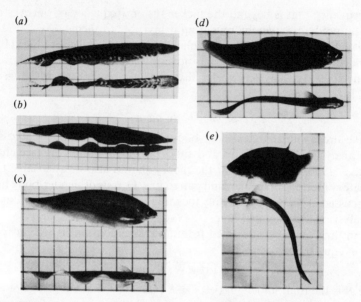

Fig. 93   Median fin propulsion in the electric-eels and knifefishes. The figure shows top and side views of *Gymnotus carapo* (*a*), *Gymnorhamphichthys hypostomus* (*b*), *Xenomystis nigra* (*c*) and *Notopterus notopterus* (*d* and *e*) swimming against a grid of 2 cm squares.

recoil effects generated during swimming; and (5) augmentation of viscous drag during swimming.

Nine species were examined. Propulsive efficiency ranged from about 0.7 to 0.9 over a speed range of about 0.2–4.5 l s^{-1}. Webb (1971b) estimated the propulsive efficiency of trout (*Salmo gairdneri*) over most of the fish's cruising range and found that values usually exceeded 0.7. At low swimming speeds however, rather low values of η_p were found (< 0.4). Electric-eels and knifefishes of a similar size to the trout studied by Webb do not show these low values of propulsive efficiency at low absolute swimming speeds. It can be argued therefore, that undulatory fin propulsion in the electric-eels and knifefishes is an adaptation to low-speed swimming with high hydromechanical efficiency.

Many fish that are propelled by oscillation of the body and caudal fin are capable of high speeds and accelerations. High speed and acceleration are necessary for the capture of prey and for escape from predators. The gymnotids and notopterids feed on slow swimming fish and aquatic insect larvae, and therefore do not require high speeds and accelerations in order to catch their prey. It is interesting to note that high swimming speeds (e.g.

escape responses) are effected by rapid bodily flexure in the subcarangiform mode in both the gymnotids and notopterids (Blake, 1983*b*).

Fig. 93 shows two gymnotids (*Gymnotus carapo* and *Gymnorhamphichthys hypostomus*; Fig. 93*a* and *b*) and two notopterids (*Xenomystis nigra* and *Notopterus notopterus*; Fig. 93*c*, *d* and *e*). *Gymnotus* and *Gymnorhamphichthys* rarely execute turning manoeuvres; by simply reversing the direction of waveform propagation on their anal fin both forms are able to swim as effectively backwards as they do forwards. In contrast, *Notopterus* shows essentially the same turning behaviour as forms that swim in the anguilliform and subcarangiform modes (see chapter 5 and Fig. 93*e*). *Notopterus* often supplements the action of its undulatory fin with a body wave during low-speed swimming.

Notopterus (and to a lesser extent *Xenomystis*) exhibit some of the morphological traits which are typical of subcarangiform fish and are designed to minimize the influence of recoil forces generated during swimming (see Lighthill, 1970, 1977). In particular, both *Notopterus* and *Xenomystis* have a good depth of section over their centre of mass. The variation in depth of section moving back from the nose is far more gradual in *Gymnotus* and *Gymnorhamphichthys* than in the notopterids (compare Fig. 93*a* and *b* with *c* and *d*).

For the symmetrical movements of the anal fin of most gymnotids the mean time-average of the recoil forces generated during locomotion will be very small, as the lateral force components produced by the fin cancel each other out on either side of the mid-line at all points except the trailing edge. In fish that swim in the anguilliform mode there is a rapid increase in the amplitude of the body wave as it passes back down the length of the fish (e.g. Gray, 1933*a*) and this gives rise to a net recoil force in addition to that already described. The anal fin of the gymnotids does not exhibit the anguilliform pattern of rapid amplitude increase; fin amplitude increases from the leading edge to a more or less constant value over most of the fin, decreasing again only close to the trailing edge. The system is essentially symmetrical.

We have noted that thrust values calculated from hydromechanical theory may be employed to infer values of the drag coefficient of the body being propelled (see chapter 5). Inferred values for the equivalent rigid-body drag of fish that swim in the anguilliform and subcarangiform modes may exceed predicted values by a factor ranging from about four (e.g. Lighthill, 1971) to nine (e.g. Alexander, 1977). It is likely that the gymnotids and other undulatory fin swimmers are not subject to such large viscous drag increments, as undulations and therefore boundary layer thinning (the supposed cause of increased viscous drag in the swimming fish) are restricted to the fins.

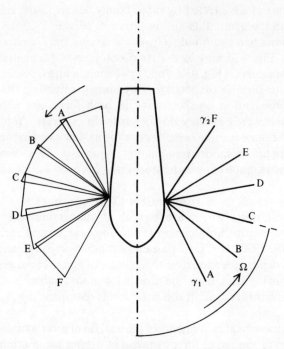

Fig. 94   Schematic diagram of fin positions during the power and recovery stroke phases of the fin-beat cycle of a fish swimming in the drag-based labriform mode. From Blake (1981*c*).

Hydromechanics of oscillatory fin propulsion

Blade-element theory of pectoral fin rowing

Pectoral fin rowing has been analysed by Blake (1979*a*, *b*, 1980*a*, 1981*a*, *c*, 1983*b*). The main features of the model of the power and recovery stroke of the fin-beat cycle and the influence of fin shape on the thrust generated during the power stroke are discussed below.

Fig. 94 shows a schematic view from above of a rowing fish. The right-hand side of the diagram depicts the fin during successive stages of the power stroke, when the fin is oriented 'broadside on' to the flow. The normal (v_n), spanwise (v_s) and resultant relative velocity of the fin during the stroke are given by

$$v_n(r, t) = \omega r - U \sin \gamma \qquad (151)$$

$$v_s(r, t) = U \cos \gamma \qquad (152)$$

$$v_{\text{res}}^2(r, t) = \omega^2 r^2 + U^2 \sin^2 \gamma - 2U\omega r \sin \gamma + U^2 \cos^2 \gamma = \omega^2 r^2 + U^2 - 2U\omega r \sin \gamma \qquad (153)$$

where ω is the angular velocity of the fin, γ is the positional angle and r is

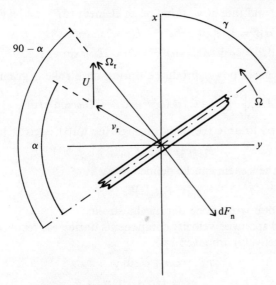

Fig. 95 A blade-element during the pectoral-fin power stroke. All notation is defined in the text. From Blake (1981c).

the distance out from the fin base. From this information we can define the hydromechanical angle of attack as

$$\tan \alpha = \omega r - U \sin \gamma / U \cos \gamma \tag{154}$$

(α and other symbols are defined below).

It is convenient to divide the fin into a series of arbitrarily defined blade-elements. Fig. 95 shows a blade-element during the power stroke. The normal force (dF_n) and thrust force produced by a blade-element are given by

$$dF_n(r, t) = \tfrac{1}{2}\rho v_{\text{res}}^2(r, t)\, dA\, C_n \tag{155}$$

$$dT(r, t) = dF_n(r, t) \sin \gamma \tag{156}$$

where dA is the area of the element and C_n is a normal force coefficient, which depends (among other things) on α. The impulse of the thrust force must be equal to the impulse of the drag force acting on the body of the fish that is being propelled, and so

$$\tfrac{1}{2}\rho U^2 S_w C_{D(b)}\, t_0 = 2 \int_0^R \int_0^{t_p} dT\, dt \tag{157}$$

where t_p is the duration of the power stroke, t_0 is the total duration of the the fin-beat cycle and $C_{D(b)}$ is the drag coefficient of the body; the factor of two arises due to the operation of two fins.

The work $(\mathrm{d}W)$ and rate of working of an element $(\mathrm{d}P)$ can be written as

$$\mathrm{d}W = \tfrac{1}{2}\rho\,\mathrm{d}A\,C_\mathrm{n}r(\omega^2 r^2 + U^2 - 2U\omega r\sin\gamma) \tag{158}$$

$$\mathrm{d}P = \tfrac{1}{2}\rho\,\mathrm{d}A\,C_\mathrm{n}\,r\omega(\omega^2 r^2 + U^2 - 2U\omega r\sin\gamma) \tag{159}$$

and therefore the mean power produced during the stroke is given by

$$\bar{P} = 1/t_\mathrm{p}\int_0^{t_\mathrm{p}}\tfrac{1}{2}\rho\,\mathrm{d}A\,C_\mathrm{n}\,r\omega(\omega^2 r^2 + U^2 - 2U\omega r\sin\gamma)\,\mathrm{d}t \tag{160}$$

The power required to drag the body through the water is given by

$$P_\mathrm{b} = \tfrac{1}{2}\rho U^3 S_\mathrm{w}C_{\mathrm{D(b)}} \tag{161}$$

The efficiency of the system can be defined as

$$\eta_\mathrm{p} = P_\mathrm{b}t_0/2W_\mathrm{T} \tag{162}$$

where W_T is the total work done during the stroke.

The normal and spanwise velocity components during the recovery stroke $(v_\mathrm{n}'$ and v_s', respectively) are given by

$$v_\mathrm{n}' = \omega'r + U\sin\gamma \tag{163}$$

$$v_\mathrm{s}' = U\cos\gamma \tag{164}$$

Denoting the spanwise, chordwise and normal components of force as $\mathrm{d}F_\mathrm{s}$, $\mathrm{d}F_\mathrm{c}$ and $\mathrm{d}F_\mathrm{n}'$, respectively, we have

$$\mathrm{d}F_\mathrm{s} = \tfrac{1}{2}\rho v_\mathrm{s}'^2 A_{(\mathrm{t})}C_\mathrm{s} = \tfrac{1}{2}\rho(U\cos\gamma)^2 A_{(\mathrm{t})}1.33\left(\frac{RU\cos\gamma}{v}\right)^{-\frac{1}{2}} \tag{165}$$

$$\mathrm{d}F_\mathrm{c} = \tfrac{1}{2}\rho(v_\mathrm{n}'\cos\Lambda)^2 A_{(\mathrm{t})}1.33\frac{c(v_\mathrm{n}'\cos\Lambda)^{-\frac{1}{2}}}{v} \tag{166}$$

$$\mathrm{d}F_\mathrm{n}' = \tfrac{1}{2}\rho v_\mathrm{n}'^2 A_{(\mathrm{t})}C_\mathrm{n} \tag{167}$$

where $A_{(\mathrm{t})}$ is the total wetted area of the fin, C_s is a frictional drag coefficient based on the spanwise velocity component for the case of a laminar boundary layer over the fin and Λ is the geometrical angle of attack of the fin.

A drag force acting in the direction of the body $(\mathrm{d}F_\mathrm{b})$ is given by

$$\mathrm{d}F_\mathrm{b} = \mathrm{d}F_\mathrm{s}\cos\gamma + \mathrm{d}F_\mathrm{c}\cos\Lambda\sin\gamma + \mathrm{d}F_\mathrm{n}'\sin\Lambda\sin\gamma \tag{168}$$

The impulse of this force over the recovery-stroke phase is given by

$$\begin{aligned}
I' &= \int_0^R\int_0^{t_\mathrm{r}}\mathrm{d}F_\mathrm{b}\mathrm{d}t \\
&= \int_0^R\int_0^{t_\mathrm{r}}(\mathrm{d}F_\mathrm{s}\cos\gamma + \mathrm{d}F_\mathrm{c}\cos\Lambda\sin\gamma + \mathrm{d}F_\mathrm{n}'\sin\Lambda\sin\gamma)\,\mathrm{d}t
\end{aligned} \tag{169}$$

where t_r is the duration of the recovery stroke. The mean power required $\bar{P}$ is given by

$$\begin{aligned}
\bar{P} &= \frac{1}{t_\mathrm{r}}\int_0^{t_\mathrm{r}}\mathrm{d}F_\mathrm{b}\mathrm{d}t \\
&= \frac{1}{t_\mathrm{r}}\int_0^{t_\mathrm{r}}(\mathrm{d}F_\mathrm{s}\,v_\mathrm{s}'\cos\gamma + \mathrm{d}F_\mathrm{c}\,v_\mathrm{n}'\cos\Lambda\sin\gamma + \mathrm{d}F_\mathrm{n}'v_\mathrm{n}'\sin\gamma\sin\Lambda)\,\mathrm{d}t
\end{aligned} \tag{170}$$

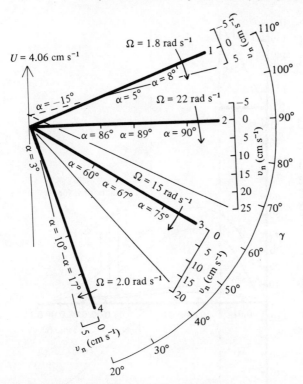

Fig. 96 Summary of the principal kinematic features of the pectoral fin power stroke of an angelfish swimming in the drag-based labriform mode. All notation is defined in the text. From Blake (1981c).

The model outlined above has been applied to the angelfish (*Pterophyllum eimekei*). Kinematic data on a single specimen ($l = 8.0$ cm) swimming at about 4.0 cm s^{-1} is summarized in Fig. 96 which shows that: (1) flow reversal and negative hydrodynamical angles of attack occur at the base of the fin at the beginning and end of the power stroke; and (2) values of v_n and α at any given instant are higher at the distal end of the fin. Due to the flow reversal, small amounts of negative (reverse) thrust are generated at the base of the fin. It is interesting to note that the outermost 40% or so of the fin area produces over 80% of the hydrodynamic thrust (see Fig. 97). Eqn 157 was employed to infer a value of $C_{D(b)} = 0.1$.

The various power components are plotted on Fig. 98. Based on eqn 162, a value of 0.26 was calculated for the efficiency of the power stroke. However, during its acceleration the fin entrains a mass of water (added mass). Taking the energy required for this into account a new value of 0.18 is calculated. The treatment of added mass in pectoral fin rowing is similar to that outlined in the following section for lift-based mechanisms. When the energy required

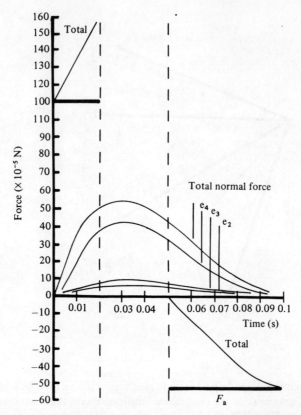

Fig. 97   Force is plotted against time for the power stroke phase of the fin-beat cycle of an angelfish. The contribution of three arbitarily defined blade-elements (e_2–e_4, the contribution of e1 was small enough to be ignored) to the total normal force, the added mass force (heavy line) and the total force are shown. From Blake (1981c).

to produce the recovery stroke is also taken into account the propulsive efficiency falls to 0.16. The impulse of the recovery stroke is about 5% of that of the power stroke.

The influence of pectoral fin shape on thrust and drag is considered by Blake (1981a). A simplified form of the power stroke model outlined above was employed to derive a shape factor B. We have

$$\tfrac{1}{2}\rho U^2 S_w C_{D(b)} t_0 = C_n B R^2 A \int_0^{t_p} \omega^2 \sin\gamma \, dt \tag{171}$$

For planforms of a given area, $B = 0.33, 0.39, 0.5, 0.6, 0.67$ for square and rectangular (irrespective of the ratio of major to minor axis), semi-elliptical, triangular (irrespective of base to height ratio), parabolic, and cubic structures,

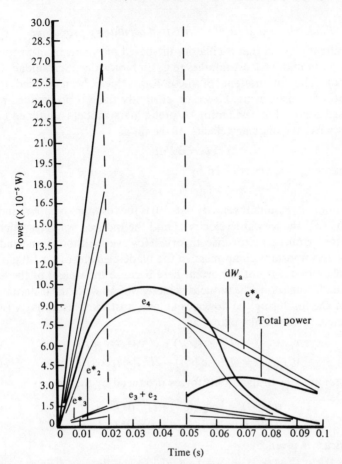

Fig. 98 The total power required to produce the hydrodynamic thrust force during the pectoral fin power stroke of an angelfish is plotted against time. The contributions to the total of the individual blade-elements (e_2, e_3, e_4), the power required to accelerate and decelerate the added mass and the contributions of the blade-elements to this (e^*_2, e^*_3, e^*_4) and the total power are indicated. From Blake (1981c).

respectively. A value of $B = 0.43$ was measured for the pectoral fins of the angelfish. Experimental work on model angelfish and pectoral fin shape combinations (Blake, 1979a, 1981a) shows that appendages of triangular planform of a given area create less interference drag over the body to which they are attached than square or rectangular ones. On the basis of the results gained from the simple hydromechanical model and the experiments, it would seem reasonable to suppose that drag-based pectoral fin swimmers would be more likely to have triangular than square or rectangular fins, and this is generally the case.

Blade-element theory of lift-based oscillatory propulsion

We have already noted that oscillatory lift-based mechanisms of propulsion are common in many teleost families (e.g. Embiotocidae, Serranidae, Chaetodontidae). The fin motions of these forms have been likened to the movements of a bird's wing. However, currently there is no detailed model of lift-based pectoral fin swimming. A preliminary model is outlined below.

We can write the angular velocity of the fin as

$$\omega = d\gamma/dt \tag{172}$$

and the flapping velocity is given by

$$v_f(r, t) = r\omega \tag{173}$$

The instantaneous resultant velocity ($v_r(r, t)$) is the resultant of $v_f(r, t)$ and two other velocities, the forward velocity U and the induced velocity which can be calculated from actuator-disc theory. However, although the induced velocity is an important component in the blade-element of bird flight (e.g. Pennycuick, 1968) it is not important here because the weight of the fish in water is small. Neglecting the induced velocity we can write the instantaneous velocity of the fin during the downstroke and upstroke, respectively (where β is the angle of the stroke plane) as

$$v_{res}(r, t) = (U^2 + v_f^2(r, t) + 2Uv_f(r, t)\cos\beta)^{\frac{1}{2}} \tag{174}$$

$$v_{res}(r, t) = (U^2 + v_f(r, t) - 2Uv_f(r, t)\cos\beta)^{\frac{1}{2}} \tag{175}$$

The instantaneous drag and lift forces produced by the fin can be written as

$$dL(r, t) = \tfrac{1}{2}\rho v_{res}^2(r, t)\, c(r)\, r^2\, dr\, C_L(r, t) \tag{176}$$

$$dD(r, t) = \tfrac{1}{2}\rho v_{res}^2(r, t)\, c(r)\, r^2\, dr\, C_D(r, t) \tag{177}$$

and the thrust is given by

$$dT(r, t) = dL(r, t)\sin\psi = \tfrac{1}{2}\rho v_{res}^2(r, t)\, c(r)\, r^2\, dr\, C_L(r, t)\sin\psi \tag{178}$$

The angle ψ is shown in Fig. 99, together with various other parameters.

The torque and power associated with the generation of the thrust force can be divided into two categories: 1. that associated with the profile drag of the fins; and 2. that associated with the inertial terms arising due to the acceleration and deceleration of the fins and their entrained added mass.

The torque associated with the profile drag (dQ_p) is

$$dQ_p = \tfrac{1}{2}\rho c(r)\, r^3\, dr\, v_{res}^2(r, t)\, C_D \tag{179}$$

and the work required over the cycle can be written as

$$W_p = \int_0^{t_0} Q_p\, d\gamma = \int_0^{t_0} \tfrac{1}{2}\rho c(r)\, r^3\, dr\, v_{res}^2(r, t)\, C_D\, d\gamma \tag{180}$$

The power associated with the profile drag is given by

$$P_p = f\int_0^{t_0} Q_p\, d\gamma = f\int_0^{t_0} \tfrac{1}{2}\rho c(r)\, r^3\, dr\, v_{res}^2(r, t)\, C_D\, d\gamma \tag{181}$$

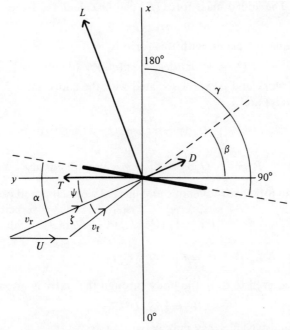

Fig. 99 Force–velocity diagram for a pectoral fin blade-element operating in the lift-based mode. All notation is defined in the text. From Blake (1983*a*).

where f is the fin-beat frequency. The inertial torque associated with the fin mass (dQ_i) can be written as

$$dQ_i = m(r)\,r^2(d^2\gamma/dt^2) \tag{182}$$

where $m(r)r^2$ is the moment of inertia of the fin. The inertial work and power can be written as

$$W_i = \int_0^{t_0} Q_i\,d\gamma = \int_0^{t_0} m(r)\,r^2(d^2\gamma/dt^2)\,d\gamma \tag{183}$$

$$P_i = f\int_0^{t_0} Q_i\,d\gamma = f\int_0^{t_0} m(r)\,r^2(d^2\gamma/dt^2)\,d\gamma \tag{184}$$

In studies of bird flight the added mass of the wings is usually neglected as it is small in comparison to the mass of the wings themselves. However, in pectoral fin swimming significant added mass terms will arise due to the relatively high density of water. By rotating the pectoral fin blade-elements about their median long axis a series of cylinders is generated. Multiplying the volume of these cylinders by the water density gives the added mass of the elements

$$M = \rho\pi(c/2)^2\,y \tag{185}$$

where y is the length of a given blade-element (measured perpendicular to the chord). The added mass force (F_a) can be calculated from

$$F_a = M(dv_f/dt) = \rho\pi(c/2)^2 y(dv_f/dt) \tag{186}$$

and the torque associated with this force is

$$Q_a = M(dv_f/dt)r = \rho\pi(c/2)^2 yr(dv_f/dt) \tag{187}$$

The total work and power associated with the added mass force (W_a and P_a, respectively) are

$$W_a = \int_0^{t_0} Q_a \, d\gamma = \int_0^{t_0} \rho\pi(c/2)^2 yr(dv_f/dt) \, d\gamma \tag{188}$$

$$P_a = f\int_0^{t_0} Q_a \, d\gamma = f\int_0^{t_0} \rho\pi(c/2)^2 yr(dv_f/dt) \, d\gamma \tag{189}$$

The overall total hydrodynamic and inertial work (W_T) and power (P_T) are

$$W_T = \int_0^{t_0} (Q_i + Q_a) + Q_p \, d\gamma \tag{190}$$

$$P_T = f\int_0^{t_0} (Q_i + Q_a) + Q_p \, d\gamma \tag{191}$$

The work required to drag the body through the water is given by

$$W_b = \tfrac{1}{2}\rho S_w U^3 C_{D(b)} t_0 \tag{192}$$

and the propulsive efficiency may be written as

$$\eta_p = H_b/2W_T \tag{193}$$

The factor of two arises due to the operation of two fins.

Blake (1983*a*) has made some preliminary observations and calculations on lift-based oscillatory propulsion in the shiner seaperch (*Cymatogaster aggregata*). The specimen examined was 9.6 cm in length and swimming at a velocity of about 23 cm s^{-1}. The pectoral fins moved through a stroke plane of about 100° (= 1.74 rad), moving from a starting position at the top of the downstroke that was about 80° from the horizontal and dipping about 20° below the horizontal at the end of the stroke. These figures are similar to those given by Pennycuick (1968) on the average arc of travel of the wings of the pigeon during forward flight. The stroke plane was about 20° (i.e. $\beta = 20°$) and corresponded well with the angle subtended by the fin base to the horizontal. The flapping frequency of the fins was approximately 5.55 Hz, giving a cycle time of about 0.18 s.

From the above information we can calculate a value for the reduced frequency parameter, $\omega\bar{c}/U$. The value calculated is close to 0.25. Low values of the parameter indicate that unsteady flow effects are not likely to be important. Values of about 0.5 have been calculated for the pigeon and locust in forward flight (Lighthill, 1974). The aspect ratio of the fins was about 4.0, and values of $C_L/C_D = 6.0$ are possible, corresponding to $C_L = 1.2$ and $C_D = 0.2$ (von Mises, 1959). However, it is unlikely that the pectoral fins of *Cymatogaster* are that 'efficient'.

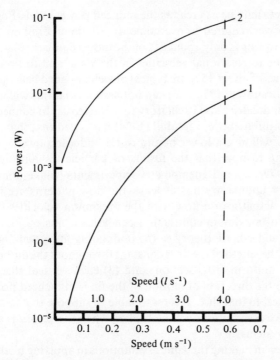

Fig. 100 The power required to overcome the frictional drag of the body (1) is compared to the total aerobic swimming power (2). From Webb (1974).

Webb (1974) calculated the swimming power of a specimen of *Cymatogaster* over a range of swimming speeds on the basis of respirometric data. Values for the total swimming power were compared to the theoretical minimum drag power to overcome the drag of the body, assuming a laminar boundary layer (see Fig. 100). Values of η_p of the order of 0.6–0.65 were calculated assuming a mechanical efficiency for the pectoral fin muscles of 0.2.

Assumptions involved in the hydromechanical models

The advantage of employing actuator-disc theory in the analysis of undulatory fin systems is that the model does not require a great deal of detailed kinematic data. The fin is reduced to the status of a 'black box' through which the water is accelerated. Ellington (1978) has pointed out that this approach can be a disadvantage if energy losses in the operating mechanism and wake are large. Four major sources of energy loss can be considered: (1) energy lost in overcoming the profile drag of the fins; (2) energy lost in tip losses; (3) energy lost in rotational velocities in the wake; and (4) extra energy required due to a non-uniform velocity distribution over the disc area.

The power required to overcome the frictional profile drag of the wings in hovering insects may be as great as the induced power (Weis-Fogh, 1973). However, the power required to overcome the frictional drag on the fins of *Synchropus* in hovering is only about 5% of the induced power (Blake, 1979*d*). Energy losses due to rotational velocities in the wake and to tip effects are not known. Values of about 15% are typical of well designed helicopter rotors and propellers (Bramwell, 1976). We can not assume that this will necessarily be the case for undulatory and oscillatory fins in teleosts. In connection with the fourth assumption above, Lighthill (1974) has noted that that a uniform induced velocity will give a lower bound to the induced power.

It is reassuring to note that the results of particle visualization studies (Blake, 1976, 1979*c*, *d*) and anemometry experiments (Blake, unpublished) indicate that, for undulatory fins at least, the flow patterns are in general agreement with actuator-disc flow with the maximum velocities coinciding with the region of maximum contraction beneath the fins.

In applying slender-body theory to the undulating fin system, we have to assume that: (1) the length of the fin is constant throughout the fin-beat cycles; (2) the fin is of uniform cross-section; and (3) each vertical slice of water perpendicular to the direction of motion of the fin is influenced primarily by the fin parts closest to that slice. It is reasonable to suppose that in most cases none of the above assumptions will be broken when the theory is applied to undulatory fin systems.

We are involved in making the same assumptions in applying blade-element theory to both drag- and lift-based mechanisms of oscillatory propulsion. We assume: (1) that the flow is steady; (2) that empirical lift and drag coefficients can be assigned to the blade-elements which can be considered as acting independently of each other; and (3) that the force coefficients are constant along the length of the fins and in time.

The assumption of steady flow is probably reasonable. In fast forward swimming the reduced frequency parameter is small and the flow dominated by the effects of free-stream convection. However, unsteady effects probably become important at low forward speeds and in hovering.

Values of C_n, C_L and C_D depend on the angle of attack of the element, Reynolds Number and the relative velocity. In many applications (e.g. bird flight studies) mean values that satisfy the force balance have been employed. Further discussion of the assumptions of blade-element theory can be found in Ellington (1978) and Blake (1979*a*, *b*, 1983*a*).

Further hydromechanical, behavioural and ecological aspects

Most fusiform, pelagic fish are capable of attaining high speeds after rapid accelerations. These attributes are essential to such forms if they are to be successful in capturing prey and escaping from predators. In contrast many undulatory and oscillatory median and/or paired fin swimmers are not capable of reaching high swimming speeds and spend much of their time in foraging at low speeds. Swimming at low speeds releases these forms from the morphological constraints associated with a streamlined body form, and many are characterized by unstreamlined profiles (e.g. Synathidae, Ostraciidae).

An advantage of low speed swimming is that highly precise manoeuvrability is possible. Many undulatory and oscillatory fin swimmers are able to turn on their own axis with no lateral translation of the body (see Fig. 89). By reversing the direction of waveform propagation on their median fins most Balistidae, Gymnotidae and Notopteridae are able to move backwards almost as well as they move forwards. In contrast, many pelagic, fusiform fish are designed for efficient steady swimming and perform relatively poorly in manoeuvre (e.g. Scombriodae).

Webb (1981) suggests that neutral buoyancy is an essential feature of slow swimming and fine manoeuvre. Negatively buoyant fish have to swim sufficiently fast to produce enough passive lift to support their weight (e.g. Magnuson, 1978). Many slow swimming forms capable of fine manoeuvre are negatively buoyant however. Lift forces are produced actively by the fins. Slow swimming and the ability to perform fine manoeuvres are essential for fish that inhabit environments such as coral reefs.

Many slow swimming coral reef fish are herbivorous, others feed on small benthic, slow swimming crustaceans (e.g. Hobson, 1974) and therefore do not require high speed and acceleration in order to ensure the capture of their prey. Many slow swimming forms that are propelled by median and/or paired fins are protected from potential predators by body armour or the ability to release toxic substances into the water. The Ostraciidae are protected by fused bony plates that totally encase the body. The Balistidae, Aluteridae and Tricanthidae are covered with rough, scute-like scales. These groups are also characterized by a sharp, retractile dorsal spine. The Tetradontidae and Diodontidae are protected by sharp spines all over their bodies. Many Balistidae, Aluteridae, Tetradontidae and Diodontidae have poisonous flesh. The boxfish *Ostracion lentiginosum* protects itself from its enemies by releasing a highly toxic substance into the water. Not all undulatory and oscillatory median and paired fin swimmers are armoured and poisonous. For example, the gymnotids and notopterids are capable of brief bouts of high-speed swimming in the anguilliform mode, and presumably escape from their predators employing rapid speed and acceleration. Most fish that are ordinarily propelled by oscillatory paired fins (labriform mode) are able to perform rapid escape responses, employing undulations of the body and caudal fin.

It is interesting to note that large mobile eyes, protective flanges over the nares and tubiferous barbules around the mouth are common in many coral reef fish that are propelled by undulatory median and/or paired fins (e.g.

Tetraodontidae, Diodontidae). Allen (1961) has shown that the eyes and nares of fish that swim in the subcarangiform and carangiform modes may act as roughness elements and cause boundary layer separation in some cases. We can speculate that it might be the case that at low swimming speeds and small size, eyes, nares and other projections do not function as roughness elements, and therefore that well developed eyes and protective nare flaps are a possible advantage of low-speed swimming. In this context it is of interest to note that eye fairings do not develop until a certain size is attained in many scombriods (Burdak, 1969, see Fig. 51).

Fish that are designed for good steady cruising performance are characterized by a high aspect ratio, forked tail fin (e.g. Carangidae, Scombriodae). In contrast, the caudal fin of many slow-swimming teleosts is of relatively low aspect ratio and spade-like in form. Caudal fin morphologies of this type are referred to as ostraciiform. On the basis of hydromechanical theory, the propulsive efficiency of the ostraciiform tail is predicted to be about 0.5 (Lighthill, in Blake, 1981b). An experimental study on *Ostracion lentiginosum* confirmed this result (Blake, 1981b). We have noted that anguilliform and carangiform swimmers are characterized by higher efficiencies than this in steady swimming (see chapter 5). In fact, values of 0.9 and above have been found for trout swimming in the subcarangiform mode (Webb, 1971b). However, the ostraciiform caudal fin is usually only employed during escape responses that involve very brief bouts of rapid swimming in 'life or death' situations where propulsive efficiency may not be of great importance.

7

Stability and control

Introduction

There have been few studies on the hydromechanical aspects of the stability and control of fish swimming. In those studies that have been done (e.g. Harris, 1936, 1937*a*, *b*, on sharks; Magnuson, 1970, 1972, 1978, on scombriods) it was implicitly assumed that the fish could be likened to a fixed-wing aircraft with similar problems of stability and control. For fish that keep their body relatively rigid during swimming and have large aerofoil-like pectoral fins forward of their centre of gravity (e.g. Scombriodae) this is probably a reasonable assumption. Most sharks and many teleosts however, are propelled by bodily undulations and have a great deal of control over their fins. Here, the rigid-body analogy may break down and a new approach be required. Further study is needed to clarify this point.

In this chapter we are concerned with the static stability of fish. The unfortunate term static stability refers to the influence of small departures from a steady straight and level course. The term dynamic stability is often used to describe large-scale movements (i.e. manoeuvres, see chapter 5). We begin by discussing the longitudinal stability of the sharks, with special reference to the role of the heterocercal tail. This is probably the most intensely studied area with respect to the stability and control of fishes to date. Both sharks and teleosts are considered in the following section which concentrates on the role of the median and paired fins in the swimming equilibrium. In this section a simple mathematical model of the stability of fish during braking is developed. The model is due to Harris (1937*a*) and remains the best currently available analysis of braking in fish.

In the final section of this chapter the equilibrium of the scombriods is discussed. The account concentrates on the negatively buoyant species that swim continuously. The role of the paired fins, caudal peduncle and keels, and the body in producing lift to counter the weight of the fish, the moment balance around the centre of gravity, changes in the relative contribution of the various lifting surfaces to the overall lift force generated in relation to speed, and morphometric variation in relation to size are all discussed.

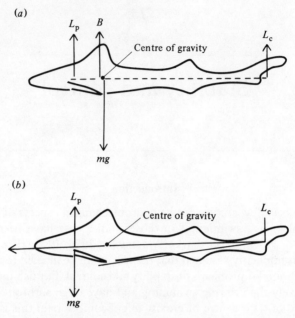

Fig. 101   Diagrams illustrating (*a*) the generally accepted view of shark equilibrium and (*b*) that suggested by Thomson (1976).

The heterocercal tail and longitudinal stability of sharks

The caudal fin of elasmobranchs is not symmetrical, the vertebral column is bent upwards at the base of the tail and lies in the dorsal (epicordal) lobe of the fin. The dorsal lobe of the fin is larger than the ventral (hypocordal) lobe. This type of caudal fin morphology is termed heterocercal. Studies on shark swimming have centred on the role that the heterocercal tail plays in the longitudinal stability of the fish.

The first serious mechanical investigation into the function of the caudal fin of sharks was conducted by Grove & Newell (1936) who constructed celluloid models of dogfish tails and tested their performance in a large tank of water. The model fins were pivoted at the end of a spindle which could be rotated. Grove & Newell found that when the model was rotated it would rise in the water column. When the rotation was stopped the model would fall. By turning the model fin upside down a downward motion could be generated. On the basis of these experiments Grove & Newell (1936) concluded that the function of the tail is to raise the tail region of the fish. In further experiments a hinge mechanism was introduced that would allow for a forward motion of the model in addition to the upward one. When the experiment was repeated the model moved upward and forward.

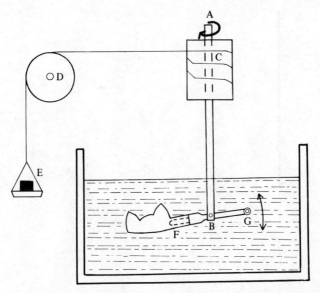

Fig. 102   Alexander's heterocercal tail experiment. A tail is fixed upside down to one end of the bar FG, which is pivoted at B to a shaft suspended from a ball race at A. The shaft rotates when weights are added to the pan E. From Alexander (1965).

Affleck (1950) repeated and extended the earlier work of Grove & Newell (1936) on model fins. Using a variety of model fin shapes, Affleck was able to show that the precise action of the fin depends on: (1) the relative size of the epichordal and hypochordal lobes; (2) its flexibility; and (3) the exact direction of the terminal vertebral axis. He also demonstrated that a symmetrical (homocercal) tail does not produce a strong upwardly directed force. These experiments confirmed the generally accepted view of the function of the heterocercal tail with respect to the longitudinal stability of the shark that had emerged in the 1930s: that the caudal fin and pectoral fins produce lift which together with a buoyancy force acting slightly anterior to the centre of gravity supports the weight of the fish (see Fig. 101a).

Alexander (1965) experimented with real caudal fins rather than models and refined the experimental approach so that the forces generated by the tails could be measured. The apparatus used by Alexander is illustrated and described in Fig. 102. In Fig. 103 the transverse speeds of the tails of the spotted dogfish (*Scyliorhinus canicula*) and the tope (*Galeorhinus galeus*) are plotted against the lift forces that they generate. From the figure it can be seen that: (1) deep-freezing has a marked effect on the properties of the tail of the tope; and (2) that the lift force generated is approximately proportional to the 1.4 power of the speed. One would expect that the lift force would be roughly proportional to the transverse velocity squared. Alexander noted that

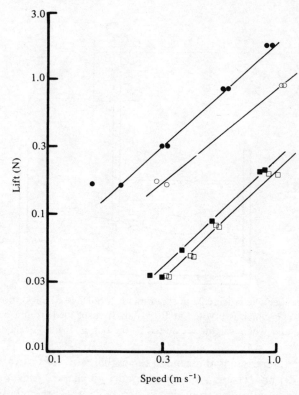

Fig. 103   Lift force is plotted against speed for the heterocercal tail of the tope (● and ○) and dogfish (■ and □). The symbols refer to unfrozen (solid symbols) and frozen (open symbols) specimens. From Alexander (1965).

the angle of attack of the hypochordal lobe of the tail decreased as the speed increased, and that therefore the lift force could be expected to increase with a power less than two.

Alexander's work was extended by Simmons (1970) who employed the same experimental approach. Simmons resolved the forces acting on the tail and discussed the significance of the direction of the resultant force produced by the tails of two quite dissimilar sharks, *Heterodontus portusjacksoni* and *Squalus megalops*. In addition, Simmons performed amputation experiments in order to assess the role of the hypochordal lobe in modifying the direction of the resultant force produced by the tail.

In *Heterodontus* the resultant force generated by the tail was inclined at an angle of about 12° to the horizontal, in *Squalus* a value of 26° was determined. These results were discussed in relation to the mode of life of the animals. Evidently, *Heterodontus* is a slow swimming fish that feeds off the bottom on molluscs and echinoderms. In contrast, *Squalus* is a fast swimming form capable of performing rapid climbing manoeuvres. The resultant force from the tail of *Squalus* passes dorsal to the centre of gravity of the fish, and therefore generates a large pitching moment, which enables the animal to rise rapidly.

Simmons (1970) reiterated the findings of Affleck (1950) with respect to the important parameters that determine the magnitude and direction of the forces generated by the heterocercal tail. In addition, he demonstrated that the hypochordal lobe of the tail acts primarily to reduce the angle of elevation of the resultant force. When the ventral lobe was amputated both types of tail were found to be strongly epibatic. Simmons suggests that through the action of the radial muscles of the caudal fin the shark may be able to operate its tail as a 'variable trim device' by adjusting the direction of the line of action of the thrust produced. This suggestion has been further discussed by Thomson (1976) and Thomson & Simanek (1978).

It has been proposed that the heterocercal tail is capable of delivering thrust that can be oriented through a large range of angles in the vertical plane and that in steady forward swimming the thrust force has a line of action that passes directly through the centre of gravity of the fish (Thomson, 1976; Thomson & Simanek, 1978). In this case there will be a net sinking component acting through the centre of gravity. Thomson & Simanek suggest that this sinking component could be offset without any epibatic or hypobatic moment by 'the planing action of the pectoral fins'. This cannot be so however, as any lift force produced by the pectoral fins will act through the centre of pressure of the fins, which lies anterior to the centre of gravity. Due to this an uncountered pitching moment will be generated. In the case of a neutrally buoyant fish no lift forces would be required from the pectoral fins for static stability. This is a purely hypothetical case. In this connection however, we can note that most sharks retain large hydrofoil-like pectoral fins and ask the question: why did the homocercal tail of the teleosts evolve when according to Thomson & Simanek the heterocercal form is adequate for their purpose?

Role of the median and paired fins in the equilibrium of fishes

Much of our knowledge in this area is due to the pioneering work of Professor J. E. Harris. Harris (1936, 1937*a*, *b*) studied the role of the fins in the swimming equilibrium of sharks and teleosts. In a particularly thorough study (Harris, 1936) he investigated the stability of a plaster model of the dogfish (*Mustelus canis*) in a wind tunnel. Prior to the work of Harris, experimental studies of swimming stability had been confined to qualitative assessments, based on fin amputation studies (e.g. Grenholm, 1923; Breder, 1926). These experiments were rather crude and not very successful, largely due to the fact

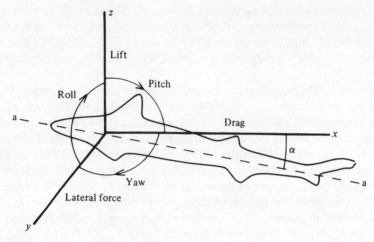

Fig. 104   Forces acting on a model of *Mustelus* in a small wind tunnel. This figure is explained in the text. From Harris (1936).

that the compensating mechanisms of the fish were so good that extreme amputations had to be performed before any effect could be seen. The account that follows is drawn largely from the work of Harris.

Harris (1936) mounted a model of *Mustelus* in a small wind tunnel. The forces acting on the model were measured with respect to the three principal axes. The x-axis corresponded to the longitudinal axis of the tunnel, and the y and z axes were measured perpendicularly (y) and vertically (z) to this (see Fig. 104). Drag, lateral force and lift forces were measured in the x, y and z directions, respectively. In addition, Harris determined the moments corresponding to rolling, pitching and yawing movements of the body ($0x$, $0y$ and $0z$, respectively, see Fig. 104). The model fish could be pivoted about its attachment to the tunnel's force balance so as to present an angle of attack to the flow.

It was assumed that no lift forces were generated when $\alpha = 0°$, and that therefore no pitching moments were produced at zero angle of attack. Harris determined the lateral force, drag and lift and the moments generated by the model at various angles of attack. Three types of experiment were performed. In the first type no fins were present on the body of the fish model. The posterior median fins (caudal, anal, and second dorsal) were attached to the model for the second type of experiment. Finally, the model was fitted with all fins. Harris noted that:

1. Lateral forces increased as α increased from a zero value and as more fins were added to the model (see Fig. 105). The form of the curves is not exactly symmetrical about $\alpha = 0°$, however the deviation is small enough to neglect.
2. Drag forces increased from a minimum obtained at small negative angles

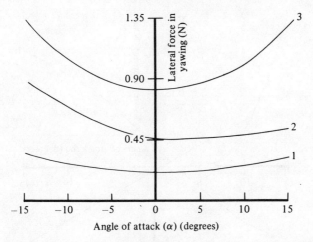

Fig. 105   Lateral force in yawing for the case of a model dogfish with no fins appended (1), posterior median fins in place (2) and all fins present (3). Data from Harris (1936).

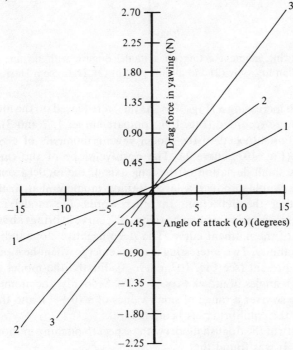

Fig. 106   Drag forces in yawing for the case of a model dogfish with no fins appended (1), posterior fins only (2) and all fins (3). Data from Harris (1936).

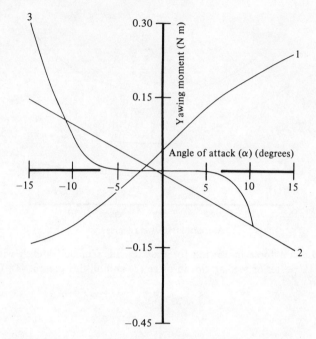

Fig. 107 Yawing moment acting on a model dogfish with no fins present (1), posterior median fins only (2) and all fins present (3). Data from Harris (1936).

of attack with increasing α (Fig. 106). As fins were placed on the model higher drag forces were recorded (Fig. 106, compare curves 1, 2 and 3).

3. With no fins attached to the model yawing moments of positive slope were found (Fig. 107, curve 1). The positive slope of the curve implies instability. A small deviation of the long axis of the model from the plane of the x-axis would produce a large turning moment that would tend to increase further the deflection. In other words, the model is unstable. Addition of the posterior median fins (Fig. 107, curve 2) brings about a change in the slope of the moment curve. The steep negative slope indicates good directional stability. Two interesting features emerge when the model is tested with all fins present (see Fig. 107, curve 3). Firstly, the model is stable at relatively high angles of attack (say, $> \pm 7°$). Secondly, the moment curve is flat or nearly so over a range of small values of α ($0 \pm 7°$), and therefore we can describe the equilibrium as being neutral.

The stability of the dogfish model with respect to pitching motions was also investigated. It was found that:

1. The curves for the lateral forces generated during pitching differed from those obtained in yaw (compare Figs 105 and 108). The curves are asym-

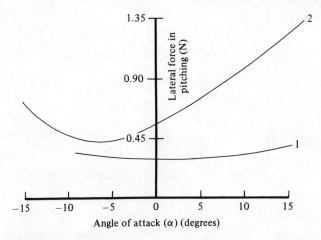

Fig. 108 Lateral forces in pitching for a model dogfish with no fins appended (1) and with all fins present (2). Data from Harris (1936).

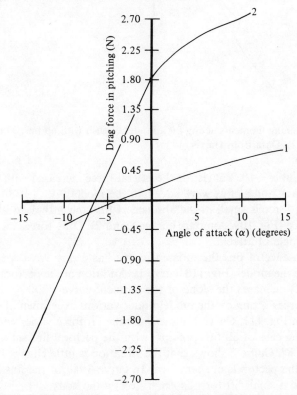

Fig. 109 Drag forces in pitching for a model dogfish with no fins appended (1) and with all fins present (2). Data from Harris (1936).

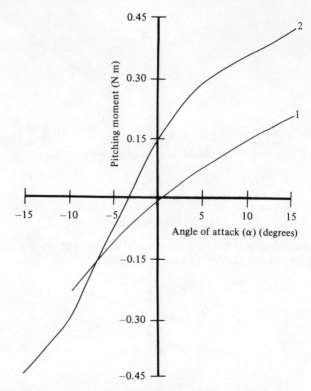

Fig. 110   Pitching moments acting on a model dogfish with no fins (1) and with all fins present (2). Data from Harris (1936).

metrical about $\alpha = 0°$. Values for the lateral force increased with increasing angle of attack and as fins were added to the model.

2. The drag force is small in the absence of fins and increases markedly in their presence (Fig. 109, compare curves 1 and 2). Drag forces increase with increasing angle of attack.

3. In the absence of fins the moment curve has a positive slope, implying instability in the model (Fig. 110, curve 1). Addition of the pectoral fins (Fig. 110, curve 2) increases the slope of the moment curve.

The lift forces acting on the model under various experimental conditions are shown in Fig. 111. Curve 1 shows how the lift force varies with angle of attack for the case of all fins present, with the pectoral fins set at an angle of attack of 8°. Curve 2 shows that the situation is little changed when fins other than the pectorals are removed. In curve 3, all of the fins have been removed and a small lift force is generated by the body.

Harris points out that the results of his wind tunnel investigations may not be entirely valid for the case of the living fish during periods of active

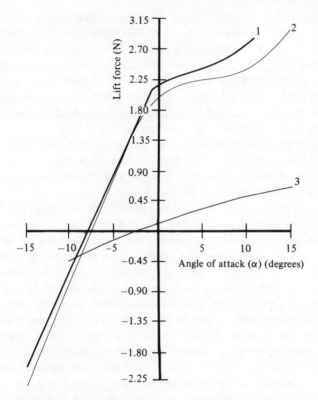

Fig. 111   Lift forces acting on a model dogfish with all fins present (1), pectorals only (2) and with no fins (3). Data from Harris (1936).

swimming. In addition, he noted that in life the fins of the dogfish are not rigid plates but are capable of a certain amount of bending. It was thought that by actively controlling the stiffness of its fins the actively swimming fish may attain a greater degree of control than that shown passively by the models. In the actively swimming dogfish it may be the case that recoil forces (generated due to asymmetries in the propulsive wave) may exceed the yawing forces produced passively by the model fish in the wind tunnel.

It is of interest to consider the role of yawing forces in a more general sense at this point. Harris (1936) noted that relative to the sharks, the teleosts have evolved a high degree of control over the movements of their fins. In the scombriods the first dorsal fin can be completely retracted into a groove in the dorsal surface of the body. Harris points out that this fin may be erected to increase the unstable moments during turning. Recently, the same turning behaviour has been described for the dolphin fish (*Coryphaena hippurus*) by Webb & Keyes (1981).

The dolphin fish is characterized by a lunate tail and large, collapsible dorsal and ventral fins. Webb & Keyes noticed that during low speed cruising activity the fish is propelled by the action of its caudal fin and the median fins are completely furled. During unsteady swimming (e.g. acceleration turns) the median fins are fully extended. In chapter 5 we noted that a high aspect ratio lunate tail is consistent with good cruising performance and that a deep section (enhanced by median fins) is required for good unsteady (turns, fast-starts) swimming performance.

It is likely that the median fins of many teleosts also play a role in preventing rolling motions of the body. In this sense they can be viewed as functioning in an analogous manner to the bilge-keels of ships (see Gadd, 1964 for an account of the function of bilge-keels).

One can view the function of the fins in fish that are propelled by bodily undulations as being governed by the need to compromise between the inversely related requirements of stability and 'controllability'. It would seem that many teleosts have solved this problem by a division of labour between the various fins in relation to the 'type' of activity performed and the level of stability required. However, to a great extent the sharks do not have this capability. It is interesting to note that the dogfish shows far more stability in yaw than it does with respect to pitch. Harris (1936) notes that with the flexibility of the body in the horizontal plane this stability can easily be overcome when turning (i.e. the fish can power its turn). However, the fish does not have the same flexibility in the vertical plane. In this case stability has been completely sacrificed in favour of controllability (i.e. the fish is inherently unstable with respect to pitching motions). It should be pointed out however that the fish has a certain amount of control over its pitching motions through the action of the pectoral fins.

In further studies Harris investigated the role of the pelvic fins in the swimming equilibrium of sharks and teleosts (Harris, 1937a, b). Evidently, the contribution to the total lift force generated by the dogfish from the pelvic fins is small. This is probably due to the influence of the 'downwash effect' of the pectoral fins. As the water passes over the pectoral fins it is deflected downward. The pectoral and pelvic fins lie in parallel planes, and so the effective angle of attack of the pelvic fins is reduced due to the influence of the pectorals. The angular reduction in the angle of attack due to this is termed the downwash angle. Amputation studies confirmed that the pelvic fins do not play an important role in the stability of the shark.

In most teleosts the pelvic fins are highly mobile. They are situated in a posterior position in the more primitive teleosts (e.g. *Esox, Lepidosteus*). Amputation experiments show that the pelvics only play a small role in the equilibrium of these forms (Harris, 1937a). In the more advanced teleosts (e.g. percoids) the pelvic fins are located more anteriorly and lie more or less below the level of the pectorals which are placed high on the sides of the body. The

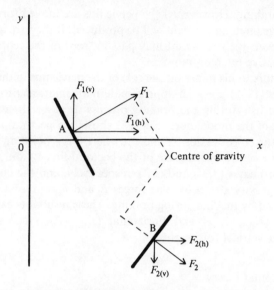

Fig. 112   This figure is explained in the text. From Harris (1937*a*).

pectoral fins function as brakes in these forms. Typically, they are placed 'broadside on' to the flow, where large drag forces (and some lift) are generated.

Harris (1937*a, b*) suggests two reasons for the forward migration of the pelvic fins in the advanced teleosts:
1. Avoidance of the downwash from the pectoral fins.
2. In braking the pectoral fins produce a drag force (Fig. 112, $F_{1(h)}$), a lift force (Fig. 112, $F_{1(v)}$) and a pitching moment which depends on the magnitude of the resultant of the drag and lift force and the distance of the moment arm from the centre of gravity of the fish. If the resultant force passes through the centre of gravity the pitching moment will be zero. This is the case in many advanced teleosts. The lift force still remains however; if this force were not neutralized the fish would rise in the water column upon braking. No upward movement can be detected however, and so we can assume that it is probable that the force is countered. The fact that when the pelvic fins are amputated the fish does rise in the water supports this argument and points to the pelvic fins as the source of the neutralizing force.

In primitive teleosts the backward position of the pelvic fins means that the downward force that they generate (Fig. 112, $F_{2(v)}$) produces a positive pitching moment which tends to depress the tail and raise the head. This tendency can be reduced or eliminated by: (1) lowering the position of the pectoral fins; (2) reducing the braking effect of the pectorals; and (3) a combination of (1) and (2). Both of the above solutions would reduce the efficiency of braking. However, if the pelvic fins are moved forward a much smaller positive pitching moment will be produced. If they are moved too far forward however, their resultant may pass in front of the centre of gravity, causing a negative pitching moment.

In an appendix to his paper on the role of the pelvic fins in the equilibrium of fish, Harris (1938) gives an approximate mathematical treatment of the equilibrium of fish during the braking process. It is worth considering the main features of the model here. Fig. 112 shows the pectoral and pelvic fin system set in the co-ordinates $0x$ and $0y$, where 0 is a point on the long axis of the fish. The centre of pressure of the pectoral fin (A) and pelvic fin (B) have the co-ordinates (f, f') and (i, i'), respectively, and the co-ordinates of the centre of gravity are $(a, 0)$. The forces F_1 and F_2 represent the resultant forces generated by the fins during braking. These resultants can be resolved into vertical $(F_{1(v)}, F_{2(v)})$ and horizontal $(F_{1(h)}, F_{2(h)})$ components. The equilibrium of vertical forces gives

$$F_{1(v)} = -F_{2(v)} \tag{194}$$

and the horizontal forces

$$F_{1(h)} + F_{2(h)} = m - \left(\frac{dU}{dt}\right) \tag{195}$$

where $-(dU/dt)$ is the instantaneous deceleration of the fish and m is the mass of the system which will include a mass of entrained water. Assuming that the drag force acting on the body passes through the centre of gravity, the equilibrium of the pitching moments about the centre of gravity gives

$$F_{1(v)}(a-f) + F_{2(v)}(a-i) = F_{1(h)}(0-f') + F_{2(h)}(0-i') \tag{196}$$

from eqns 194 and 195, we can write

$$F_{1(v)}(i-f) = F_{1(h)}(0-f') + F_{2(h)}(0-i') \tag{197}$$

If the pectoral fins are ventrally placed (as they are in primitive teleosts) we can assume that $f' = i$ and

$$F_{1(v)}(i-f) = (F_{1(h)} + F_{2(h)})(0-i') \tag{198}$$

From eqn 198 we can conclude that: (1) the pelvic fins must lie posterior to the pectoral fins (as all terms are positive); and (2) since $F_{1(v)}$ is small compared to $(F_{1(h)} + F_{2(h)})$, $(i-f)$ must be large in comparison with $(0-i')$ and therefore the distance from the pectoral to the pelvic fins will be large (e.g. trout, pike).

In the case of the advanced teleosts (e.g. percoids) the pectoral fins are as high as the centre of gravity (i.e. $0 = f'$) and from eqn 198

$$F_{1(v)}(i-f) = F_{2(h)}(0-i') \tag{199}$$

Comparing eqn 198 with eqn 199 we note that:
1. The right-hand side of eqn 199 is $[F_{2(h)}/(F_{1(h)} + F_{2(h)})]$ of that of eqn 198.
2. Since $F_{2(h)}$ is small relative to $F_{1(h)}$, $[F_{2(h)}/(F_{1(h)} + F_{2(h)})]$ is small, and $(i-f)$

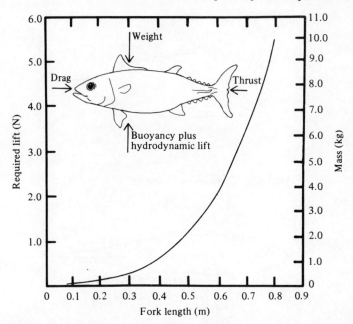

Fig. 113   Fork length is plotted against lift force required to support the weight for *Euthynnus affinis*. The inset shows the forces acting on the fish. Based on Magnuson (1970).

must be much greater than it is in eqn 198. The value of i is decreased, and the pelvic fins are closer to the pectorals in position (e.g. percoids).

Hydrostatic equilibrium of the scombriods

Many scombriods lack a swimbladder and are negatively buoyant (e.g. Magnuson, 1966). It has long been thought that they swim continuously in order to generate lift to counter their excess weight over buoyancy (e.g. Denton, 1961; Alexander, 1967). In fact, five species are known to swim continuously: the Atlantic mackerel (*Scomber scombrus*), Pacific bonito (*Sarda chiliensis*), skipjack tuna (*Katsuwonus pelamis*), wavyback tuna (*Euthynnus affinis*) and the yellowfin tuna (*Thunnus albacores*) (Magnuson, 1970). Magnuson (1970, 1972) has made a series of detailed studies on the hydrostatic equilibrium and general swimming performance of a variety of scombriod and xiphoid fish. Most of the following account is drawn from his work.

Magnuson (1970) investigated the equilibrium of *E. affinis*, employing morphometric measurements, determinations of body density and observations of swimming behaviour of wild and captive fish. He was able to show

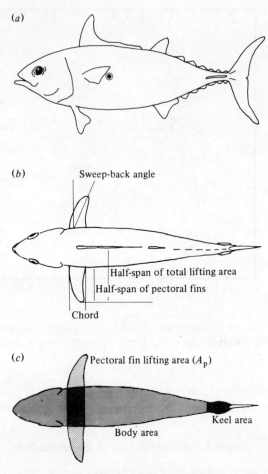

Fig. 114   The position of the centre of gravity in *Euthynnus affinis* (*a*) and some important morphometric parameters (*b* and *c*). From Magnuson (1970).

that the fish could not obtain neutral buoyancy by depositing low density compounds such as fats, and that wild fish are markedly negatively buoyant with a specific gravity of about 1.09. Magnuson assumed that the weight of the fish must be countered by lift forces produced by the pectoral fins, which appear very much like thin, cambered aerofoils. The lift force required to support the weight of *E. affinis* is plotted in Fig. 113, the forces acting on the fish are also shown (inset).

Employing much of the same fluid mechanics outlined in chapter 1, Magnuson found that:
1. The weight of water of *E. affinis* is related to fork length (length from the

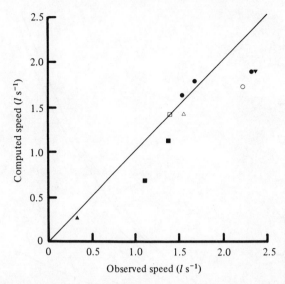

Fig. 115 Computed minimum swimming speed (from eqn 201) is plotted against observed speeds for wahoo (▲), bigeye tuna (■), yellowfish tuna (□), Pacific bonito (△), skipjack tuna (●), kawakawa (○), and bullet mackerel (▼). From Magnuson (1972).

snout to the point at which the lobes of the caudal fin meet) by the power equation $0.596\ l^{3.14}$ (dynes).

2. The lift from the pectoral fins acts through a point (centre of pressure) which is anterior to the centre of gravity of the fish.

3. The total lifting area of the pectoral fins is related to fork length by the expression $0.0134\ l^{2.13}$ (cm²).

4. The sweep-back angle of the pectoral fins increases with increasing swimming speed (i.e. the fins are extended less at high speeds).

5. The total lift, surface loading (lift per unit area) and lift coefficient of the pectoral fins varied with speed.

Magnuson proposed that the pitching moment generated by the pectoral fins about the centre of gravity is countered by a moment produced by the keel (caudal peduncle and its associated appendages). This suggestion was first made by Gregory (1928). When the pectoral fins are extended in their 'most efficient posture' they probably carry about 80% of the fish's weight, and therefore only about 20% of the weight is supported by the keel. Magnuson calculated that this corresponds to a lift coefficient of about 0.8 for the pectoral fins and keel. However, the surface loading is calculated to be higher for the keel.

The increasing sweep-back angle with increasing speed implies that the

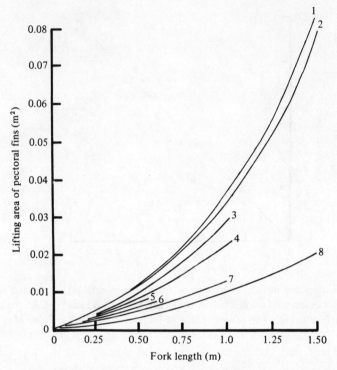

Fig. 116   Fork length is plotted against pectoral-fin lifting area for *Thunnus pelamis* (1), *T. albacores* (2), *Euthynnus affinis* (3), *Katsuwonus pelamis* (4), *Scomber japonicus* (5), *Auxis rochei* (6), *Sarda chiliensis* (7) and *Acanthocybium solandri* (8). From Magnuson (1972).

pectoral fins are producing less lift at higher swimming speeds. At very high speeds the pectoral fins are furled and presumably produce no lift at all. Magnuson suggests that at high swimming speeds up to 70% of the lift force required is generated by the body. Although values of the lift coefficient calculated for the body are relatively low (0.1–0.4) the area generating the force is large. Some of the measurements made by Magnuson and morphometric definitions that he employs are indicated in Fig. 114.

Like fixed-wing aircraft, the scombrioids must progress at a velocity that is equal to or greater than the minimum speed required to generate the lift force needed to support their weight. Magnuson was able to calculate this minimum speed (U_{min}) from

$$U_{min} = \left[\frac{L_{tot}}{\rho/2(C_{Lp} A_p + C_{Lk} A_k)} \right]^{\frac{1}{2}} \tag{200}$$

where L_{tot} is the total lift force required to support the weight, C_{Lp} is the lift coefficient of the pectoral fins, C_{Lk} is the lift coefficient of the keel, A_p is the effective pectoral fin area, A_k is the area of the keel. Values of U_{min} predicted from eqn 200 were compared to the observed swimming speeds of fish of given length for *E. affinis* and six other species (Magnuson, 1972). Computed speeds are plotted against observed swimming speeds in Fig. 115 which shows that the two speeds are highly correlated. In Fig. 116 fork length is plotted against the lifting area of the pectoral fins for the seven species investigated by Magnuson. It is reasonable to hypothesize that those species with large pectoral fins (and therefore a greater lifting force relative to fork length) should have a lower minimum swimming speed than those with small pectoral fins. Comparing Figs 115 and 116 we see that this is generally the case.

8

Swimming strategy

Introduction

Until recently little thought was given to the possibility that for reasons of energetic advantage or range increase, fish that swim in any given propulsive mode may opt for a particular swimming strategy. For example, any given species has the option to cover a given distance by swimming steadily at a constant speed or by alternating periods of active swimming with periods of passive gliding (burst-and-glide swimming). In fact it can be shown that burst-and-glide swimming can be energetically more efficient than steady swimming and can also lead to substantial increases in the maximum range possible on a given energy store. This and other swimming strategies are the subject of this chapter.

Firstly, the concept of optimal cruising speed is discussed for neutrally buoyant teleosts. A simple hydromechanical model is outlined that leads to expressions that describe the most energetically favourable velocity for steady swimming over long distances (e.g. migrations, etc.). Optimal swimming speed models based on criteria other than energetic efficiency of swimming and range (e.g. growth rate) are also considered.

A hydromechanical model of burst-and-glide swimming is outlined next. The ratio of the energy requirement per unit distance travelled in burst-and-glide swimming to that in the steady mode indicates that burst-and-glide swimming can be up to 4–6 times more energetically efficient than steady swimming. In the same section, recent experimental determinations of the drag of fish swimming in both modes are discussed.

A form of intermittent swimming is also possible for negatively buoyant fish such as the scombriods. Negatively buoyant fish are able to glide downward at an angle to the horizontal and return to their original depth by active swimming. It is shown that this form of intermittent swimming is also associated with large energy savings and large potential range increases on a given energy store beyond that of steady swimming at a constant velocity and level.

Both neutrally and negatively buoyant fish can potentially make use of

158

tidal-stream transport. In tidal-stream transport the fish only swims when the tide's direction coincides with its own. The fish rests on the bottom when the tide is unfavourable for transport. A hydromechanical model of the process is outlined. The model sets out to describe the energetic significance of tidal-stream transport by comparing the energy requirements for the process with those of constant steady swimming at constant speed irrespective of the direction of the tide relative to the fish. Again, large potential savings in energy and large range increases are indicated.

Fish schooling has long fascinated biologists and a variety of explanations have been suggested for the function of schools. Breder (1967) discusses possible social and genetic functions and Brock & Riffenburgh (1960) consider the possible advantages of schools with respect to the predatory–prey relationships of fish. Recently, possible hydrodynamic aspects of schooling have been investigated both theoretically and experimentally. Hydromechanical models of schooling make predictions concerning optimal spacing requirements for maximum energetic advantage. Experimental studies have centred on attempts to establish whether or not fish swim in the positions predicted by the theory.

Porpoising is a striking habit of dolphins and their close relatives. Dolphins are essentially confined to the surface layers of the oceans because they have to breathe air at intervals. It is known that substantial drag increments occur on streamlined bodies that move close to the surface of a body of water, making swimming there energetically expensive. It would seem that one way of reducing the energy required to cover a given distance close to the surface is to porpoise. The energetic advantages of porpoising over steady swimming close to the surface are discussed.

The final swimming strategy considered here (although doubtless other possibilities remain to be identified) is the use many demersal, negatively buoyant fish make of the ground effect. In this application the ground effect refers to the reduction in the induced power required for hovering and slow forward movement in the influence of the ground, relative to that out of ground effect.

For the sake of convenience and clarity possible swimming strategies are discussed separately here. It is important to bear in mind however, that few of them are potentially mutually exclusive and that in the wild fish may employ combinations of swimming strategies for maximum effect. It is also important to note at this point that most of the mechanisms that are discussed here have not been experimentally investigated and are essentially unproven.

Optimal cruising speed

The concept of optimum cruising speed becomes important when considering the long-range movements of fish such as feeding and spawning migrations.

Studies on optimum cruising speed to this point have been largely theoretical and aimed at determining one or more of the following: (1) optimum speed with respect to maximum range; (2) optimum speed for minimizing the energy required to cross a given distance; and (3) optimum speed with respect to feeding efficiency and growth.

Weihs (1973b) approaches the first two issues. The total available energy of a fish can be written as

$$W_{tot} = (P_s + P_m)\,t \tag{201}$$

where P_s and P_m are the swimming power and the rate at which energy is expended on basal metabolism, respectively, and t is the swimming time. Swimming thrust is equal to drag, and so

$$T = \tfrac{1}{2}\rho S_w U^2 C_{D(b)} \tag{202}$$

Thrust can be related to swimming power

$$\eta_p P_s = TU \tag{203}$$

Propulsive efficiency is in turn related to U by a constant (χ) which must be empirically determined

$$\eta_p = \chi U \tag{204}$$

Substituting eqns 202 and 204 into 203 gives

$$P_s = \frac{\tfrac{1}{2}\rho S_w C_{D(b)} U^2}{\chi} \tag{205}$$

and eqn 201 can be written as

$$W_{tot} = \left(\frac{\tfrac{1}{2}\rho S_w C_{D(b)} U^2}{\chi} + P_m\right) t \tag{206}$$

The distance crossed in time t (s) is given by Ut. By eliminating t from eqn 206 we can write

$$s = \frac{W_{tot} U}{(\tfrac{1}{2}\rho S_w C_{D(b)} U^2/\chi) + P_m} \tag{207}$$

It can be seen from eqn 207 that s has a maximum value corresponding to some positive value of U. Differentiating s with respect to U gives

$$U^2{}_{max} = \frac{\chi P_m}{\tfrac{1}{2}\rho S_w C_{D(b)}} \tag{208}$$

which is equivalent to

$$P_{s(max)} = P_m \tag{209}$$

Eqn 209 says that the maximum range possible with a given energy store is obtained when the swimming and basal metabolic power are equal. Both swimming power and basal metabolic power can be obtained experimentally through respirometry. Weihs found that eqn 209 is satisfied when U is equal to about $1.0\ l\,s^{-1}$ (see Fig. 117).

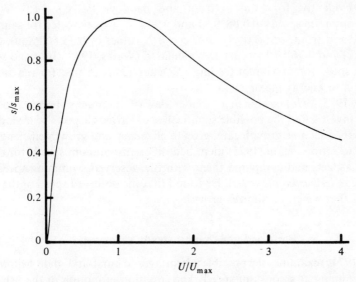

Fig. 117   Ratio of range to maximum range is plotted against normalized velocity for constant-speed swimming. From Weihs (1973*b*).

In order to assess the magnitude of the gains involved in swimming at the optimum swimming speed we can write

$$s = \frac{U_{\max} W_{\text{tot}}}{(\frac{1}{2}\rho S_w C_{D(b)} {}^2 U^2{}_{\max}/\chi) + P_m}$$

(210)

where speeds other than $U_{\max}$ are written in terms of $U_{\max}$ as $U = \sigma U_{\max}$. The maximum range ($s_{\max}$) is obtained by setting $\sigma = 1$. Applying eqn 208 we have

$$\frac{s}{s_{\max}} = \frac{2\sigma}{(1+\sigma^2)}$$

(211)

$s/s_{\max}$ is plotted against $U/U_{\max}$ in Fig. 117 which shows that the range obtainable is within 20% of the maximum for speeds one-half to twice the optimum. In a further study, Weihs (1977) investigated the effect of size on sustained swimming speeds and found that $U_{\max}$ varied rather slowly with size.

Weihs (1975) equates a swimming energetics model of the type outlined above with a model of the energy intake during feeding and solves for an optimum speed with respect to feeding efficiency. The optimum velocity for maximum feeding efficiency is given by

$$\frac{P_m}{P_s} = 2 - \epsilon$$

(212)

where ϵ is an empirically obtained constant (from the efficiency versus velocity curve). Evidently, for salmon, trout and haddock $0.7 < \epsilon < 1.0$. When $\epsilon = 1.0$, optimum speeds of 0.88, 0.91 and 0.94 l s^{-1} are calculated for salmon, haddock and trout, respectively. For $\epsilon = 0.7$, values of 0.73 (salmon), 0.72 (haddock) and 0.76 l s^{-1} (trout) are calculated (Weihs, 1975). It would seem that the speed for maximum feeding efficiency tends to that for maximum range and/or energy maximization as $\epsilon \to 1.0$.

Ware (1975, 1978) has taken a broader view of optimum swimming speeds and has investigated the possible significance of various aspects of biological performance such as growth rate, growth efficiency and growth efficiency in relation to ration. Ware (1975) identified different optimum speeds for each of these aspects and compared them with the observed swimming speed of small bleak (*Alburnus alburnus*). He found that the observed speed of the fish matched that which maximizes growth.

Burst-and-glide versus steady swimming

Weihs (1974a) examines the possible advantages of burst-and-glide swimming over swimming at a constant speed. The equation of motion of the fish can be written as

$$T = m(\mathrm{d}U/\mathrm{d}t) + \kappa(\tfrac{1}{2}\rho S_{\mathrm{w}} C_{\mathrm{D(b)}}) U^2 \qquad (213)$$

where κ is a factor by which the swimming drag can be expected to exceed the rigid-body equivalent drag (see chapter 5). Weihs set κ at 3 and, assuming a linear relation between propulsive efficiency and swimming speed (i.e. $\eta_{\mathrm{p}} = \chi U$), wrote the energy required to cross a given distance s in time t as

$$W = \int_0^t (1/\eta_{\mathrm{p}}) T U \,\mathrm{d}t \qquad (214)$$

Travelling at a constant velocity U_{c} ($U_{\mathrm{c}} = s/t$) we have

$$W_{\mathrm{c}} = (1/\eta_{\mathrm{p}}) T_{\mathrm{c}} U_{\mathrm{c}} (\tfrac{1}{2}\rho S_{\mathrm{w}} C_{\mathrm{D(b)}}) = (T_{\mathrm{c}}/\eta_{\mathrm{p}}) s \qquad (215)$$

Defining the energy expenditure per unit distance $W^* = W/s$, we obtain

$$W^*_{\mathrm{c}} = \frac{T_{\mathrm{c}}}{\eta_{\mathrm{p}}} \qquad (216)$$

For constant speed the acceleration term in eqn 213 disappears and, applying the linear dependence of propulsive efficiency on U

$$W^*_{\mathrm{c}} = \frac{\kappa(\tfrac{1}{2}\rho S_{\mathrm{w}} C_{\mathrm{D(b)}})}{\chi} U_{\mathrm{c}} \qquad (217)$$

For burst-and-glide swimming the fish produces thrust for a period and then glides passively. Taking t_1 to be the time for the acceleration phase, the energy required per cycle can be written as

$$W_{\mathrm{g}} = \int_0^{t_1} (T/\chi U) U \,\mathrm{d}t \qquad (218)$$

Weihs writes the thrust in terms of an equivalent steady swimming speed U',

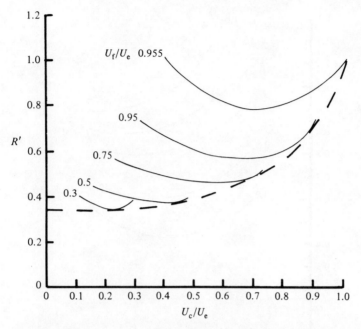

Fig. 118 Ratio of energy per unit distance travelled in burst-and-glide swimming to that for steady swimming (R') is plotted against normalized velocity (U_f = highest velocity obtained in a burst-and-glide cycle). From Weihs (1974a).

and so eqn 218 can be written as

$$W_g = \int_0^{t_1} \frac{\kappa(\frac{1}{2}\rho S_w C_{D(b)})\, U'^2}{\chi}\, dt = \frac{\kappa(\frac{1}{2}\rho S_w C_{D(b)})}{\chi} \int_0^{t_1} U'^2\, dt \tag{219}$$

Dividing by the distance travelled during a cycle gives the energy required per unit distance

$$W^*_g = \frac{W_g}{s_1 + s_2} = \frac{\kappa(\frac{1}{2}\rho S_w C_{D(b)})}{\chi(s_1 + s_2)} \int_0^{t_1} U'^2\, dt \tag{220}$$

where s_1 and s_2 are the distances traversed in the acceleration and gliding phases, respectively. Weihs then defines the ratio of the energies per unit distance traversed R' as

$$R' = \frac{t_1 + t_2}{(s_1 + s_2)^2} \int_0^{t_1} U'^2\, dt \tag{221}$$

When R' is less than one, burst-and-glide swimming will be advantageous relative to swimming at the same average constant swimming speed. In Fig. 118 R' is plotted against average velocity (normalized relative to the maximum sustained swimming speed, U_e) for trout with $\kappa = 3$. We note:
1. For $U_c/U_e = 1$, $R' = 1$ and there is no advantage to burst-and-glide swimming over constant speed swimming.

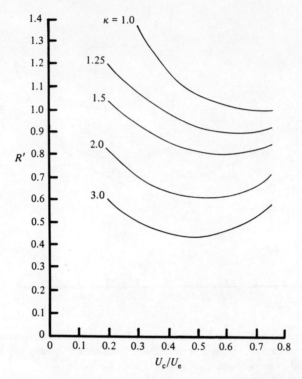

Fig. 119   Energy ratio of burst-and-glide swimming to constant-swimming speed is plotted against normalized velocity for different values of the drag augmentation factor κ. From Weihs (1974a).

2. For $U_c/U_e < 1$, $R' < 1$ and burst-and-glide swimming is energetically favourable relative to steady swimming.
3. Curves are drawn that correspond to various combinations of initial and final velocities (U_f = final velocity at the termination of a gliding phase). It can be seen that the value of U_f/U_e should not exceed the average velocity for the cycle by very much because as $U_f/U_e \to 1.0$, so does R'.
4. The smaller the required average velocity, the larger the benefit of burst-and-glide swimming.

The theoretical minimum value of R' ranges from 1 at $U_c/U_e = 1$ to $1/\kappa$ as the average velocity for the cycle tends to zero. The influence of κ on R' at various normalized swimming speeds for $U_f/U_e = 0.75$ is indicated in Fig. 119. Small values of κ correspond to large values of R' and, therefore, to small energy gains. It is worth pointing out that the above analysis is dependent upon κ being greater than one for any positive results. This means that fish to which the rigid-body analogy applies (see chapter 6) could not attain any

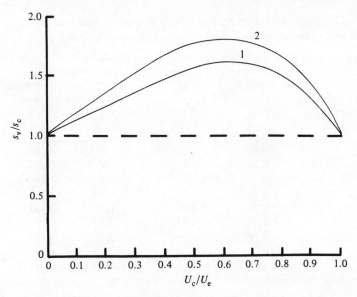

Fig. 120 Range in burst-and-glide mode of swimming relative to that in steady swimming is plotted against normalized velocity for salmon (curve 1) and haddock (curve 2). From Weihs (1974a).

advantage from burst-and-glide swimming. This is consistent with the fact that such fish do not swim in the burst-and-glide mode.

Weihs (1974a) extends his analysis to look at energy considerations, such as the range available on a given energy store, S. For constant speed swimming

$$S = (P_c + P_m) t_c \qquad (222)$$

where P_c is the power required to swim at U_c and t_c is the time available before the energy store is used up. For burst-and-glide swimming

$$S = (P_g + P_m) t_v \qquad (223)$$

where P_g is the average rate of energy expenditure for burst-and-glide swimming (at an average velocity equal to U_c) and t_v is the time available before the depletion of S. Weihs (1974a) shows that

$$t_v/t_c = [1 + (P_m/P_c)]/[1 + (P_m/P_g)] (P_c/P_g) \qquad (224)$$

and $$s_v/s_c = [1 + (P_m/P_c)]/[R + (P_m/P_g)] \qquad (225)$$

where s_c and s_v are the distances travelled under each mode.

Employing calculated values of R (Figs 118 and 119) and taking values of P_m/P_c from the metabolic literature, Weihs was able to plot s_v/s_c against U_c/U_e for salmon (*Oncorhynchus nerka*) and haddock (*Melanogrammus aeglefinus*) (Fig. 120). Apart from the limit of s_v/s_c at $U_c = 0$ and $U_c = U_e$ it can be seen

Table 2   *Comparison of drag during steady swimming with that in burst-and-glide swimming for a 0.30 m cod. From Videler (1981)*

	U (m s^{-1})	Drag (N)	Acceleration (m s^{-2})	Drag coefficient
Steady swimming	0.39	0.04	−0.05	0.019
	0.44	0.13	+0.18	0.049
	0.59	0.18	–	0.037
	0.60	0.20	–	0.040
	0.76	0.41	–	0.051
	0.85	0.64	–	0.064
Kick phase	0.76	2.51	3.10	0.310
	0.95	4.50	6.20	0.360
	1.08	2.81	3.10	0.170
Glide phase	0.76	0.12	−0.41	0.015
	0.95	0.14	−0.46	0.011
	1.08	0.18	−0.60	0.011

that burst-and-glide swimming is highly advantageous to these fish. Fig. 120 indicates that range increases of 80 and 50% are possible for the salmon and haddock in the burst-and-glide mode.

In a recent study of cod swimming, Videler (1981) uses Lighthill's elongated-body theory (see chapter 5 for a discussion of elongated-body theory) to calculate the force and power requirements of both steady and burst-and-glide swimming. Contrary to the prediction of the Weihs model, the cod (*Gadus morhua, l* = 0.3 m) did not perform burst-and-glide swimming at velocities below the maximum cruising speed. In fact, the animal swam in the burst-and-glide mode only at higher forward speeds. Videler (1981) points out that at low forward speeds energy saving is probably not important as the aerobic red muscles can deliver the required power for swimming indefinitely without using an energy store (see chapter 2) and so there is no requirement for burst-and-glide swimming. At higher speeds however energy saving becomes important and burst-and-glide swimming can be expected to occur. This argument can also be applied to the previous discussion of optimum cruising speeds.

Videler (1981) also calculates the drag coefficient of a specimen of *Gadus* in steady and burst-and-glide swimming. It is interesting to note that values of $C_{D(b)}$ for steady swimming typically exceed those for gliding by a factor ranging from four to six, and that values for the 'kick' phase of burst-and-glide swimming are far higher still (see Table 2). In his model Weihs does not take differences in $C_{D(b)}$ into account. In addition, it is assumed that the propulsive efficiency of burst-and-glide and steady swimming are the same. It has been

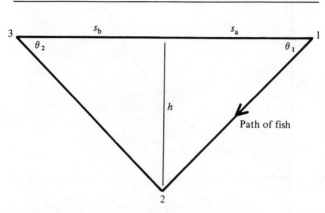

Fig. 121   Glide path of a negatively buoyant fish. This figure is explained in the text. Based on Weihs (1974*b*).

shown that steady swimming is associated with a higher propulsive efficiency than swimming in the burst-and-glide mode however (e.g. McCutcheon, 1977). When the differences are taken into account the energy savings predicted by the model are reduced from about 50% to 30% (Blake, 1981*b*, 1982*b*).

Cruising speed of negatively buoyant fish

Weihs (1973*c*) considers the relative energetic advantages of a negatively buoyant fish (e.g. Selachii, Scombriodae) swimming in two different modes: (1) steady swimming at a constant speed and level; and (2) intermittent swimming, where the fish glides downward at a constant velocity at an angle to the horizontal and then returns to its original level by active swimming at some other angle.

We begin by considering a submerged fish of weight *mg*. The weight is countered by lift forces that are generated by the fins (see chapter 7). The drag forces acting on the fish are balanced by the thrust. Weihs assumes that when the fish arrives at a point 1 (see Fig. 121) it stops all swimming movements and glides downward at an angle θ_1. The forces acting on the fish are

$$(mg)\sin\theta_1 = D \tag{226}$$

$$(mg)\cos\theta_1 = L \tag{227}$$

During this gliding phase the fish covers a horizontal distance (s_a)

$$s_a = h\cot\theta_1 \tag{228}$$

When the fish reaches the end of its glide (Point 2, Fig. 121) it starts to swim

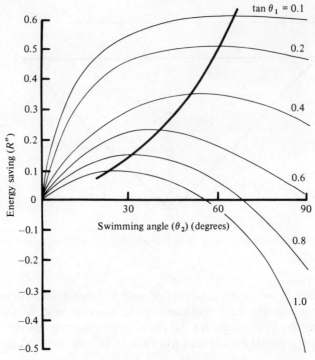

Fig. 122   Non-dimensional energy saving (R'') versus swimming angle (θ_2) for various glide angles (θ_1), $\kappa = 3$. Maximum savings are indicated by the heavy line. From Weihs (1974*b*).

upward at some other angle θ_2. The forces are

$$T = \kappa D + (mg)\sin\theta_2 \tag{229}$$

$$L = (mg)\cos\theta_2 \tag{230}$$

Eqn 230 can be written as

$$T = \kappa D\left(1 + \frac{1}{\kappa}\frac{\sin\theta_2}{\sin\theta_1}\right) \tag{231}$$

The distance travelled during this phase is

$$s_b = h\cot\theta_2 \tag{232}$$

and therefore the total distance travelled is

$$s_{tot} = s_q + s_b = h(\cot\theta_1 + \cot\theta_2) \tag{233}$$

The energy required to cross s_{tot} in steady swimming is given by

$$W_c = (1/\eta_m)\kappa Ds_{tot} \tag{234}$$

where η_m is the mechanical efficiency of the muscles (see chapter 2). For intermittent swimming

$$W_{int} = (1/\eta_m)\,T(s_b{}^2 + h^2)^{\frac{1}{2}} \tag{235}$$

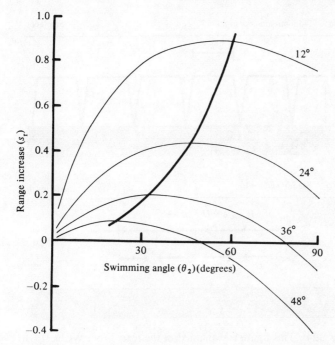

Fig. 123 Non-dimensional range increase (s_r) is plotted against swimming angle (θ_2) for various glide angles (θ_1). From Weihs (1974*b*).

From the above, a non-dimensional relative energy saving R'' can be defined as

$$R'' = \frac{W_c - W_{int}}{W_c} \qquad (236)$$

Values of R'' are plotted against θ_2 in Fig. 122 for various values of the glide angle (θ_1), with κ set at 3. We can note that:

1. As the glide angle increases the maximum value of R'' that is possible falls. When glide angles exceed about 30° negative values of R'' occur for high values of θ_2.

2. Savings of more than 50% are possible for glide angles of the order of 10°.

Weihs (1973*c*) also considers the possible range increases for intermittent swimming with downward gliding over steady swimming at constant speed and level. The non-dimensional range increase (s_r) is plotted against θ_2 for various glide angles in Fig. 123. For $\kappa = 3$ and $\theta_1 = 12°$, an increase in range of over 90% is found for the intermittent swimming mode.

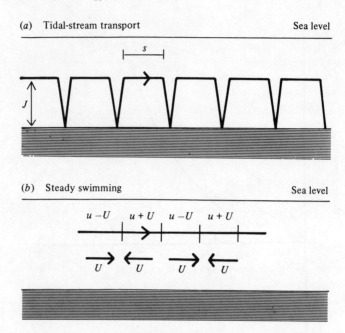

Fig. 124   This figure is explained in the text. From Weihs (1978).

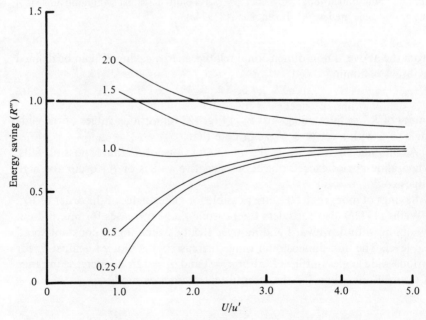

Fig. 125   Non-dimensional energy saving in tidal-stream transport is plotted against a non-dimensional velocity index (forward velocity/tide velocity) for the case of different values of the ratio of the optimum swimming speed (U_o in the absence of tides) to U. From Weihs (1978).

Tidal-stream transport

It has been suggested that fish such as plaice (*Pleuronectes platessa*) and sole (*Solea solea*) can use the tides to aid migrations, by resting on the seabed when the tide is acting against them and swimming when the tide is favourable (Greer-Walker, Harden-Jones & Arnold, 1978). Weihs (1978) has constructed a hydromechanical model of tidal-stream transport. The model describes the energetic significance of the process by comparing its energy requirements with those of steady swimming at a constant velocity.

The propulsive work required for steady swimming for a tidal cycle can be written as

$$W_c = \tfrac{1}{2}\rho S_w C_{D(b)}(U-u')^2 s + \tfrac{1}{2}\rho S_w C_{D(b)}(U+u')^2 s \qquad (237)$$

where u' is the tidal speed and s is the total distance swum. To cross the same distance ($2s$) in tidal transport the work is

$$W_{tp} = \rho S_w C_{D(b)}(U-u')^2 s + \rho S_w C_{D(b)} U^2 J \qquad (238)$$

where J is the height of the fish above the bottom (see Fig. 124).

The work required for maintenance at the standard rate (W_m) is also included in the model. Weihs notes that the time to cross the distance $2s$ in tidal-stream transport is twice that which would be spent in steady swimming. Taking this into account Weihs compares the total rate of energy expenditure in the two modes (as a ratio, R'''). Values of R''' corresponding to given values of the ratio of the optimum swimming speed to u' can then be plotted against U/u' for a given value of J/s (e.g. Fig. 125).

Fig. 125 shows that: (1) tidal-stream transport is energetically more efficient than steady swimming for values of U_o/U (where U_o is the optimum swimming speed) < 1.5 for $U/u' > 1.0$; and (2) for any given value of U_o/U that is less than about 1.5, the advantage of tidal-stream transport over steady swimming decreases for increasing values of U/u'. Weihs (1978) predicts savings in the energy cost per unit distance of over 90% and 40% for juvenile and adult fish (such as plaice and sole), respectively. It is interesting to speculate that under certain circumstances that time may be of greater importance than energy saving and that because of this certain species may favour steady swimming as opposed to tidal-stream transport.

Hydrodynamic aspects of schooling

Weihs (1973*b*, 1974*b*) has developed a two-dimensional inviscid model of schooling that can be employed in making predictions on the optimal formation and spacing of fish in schools for maximum energetic advantage. The model assumes that each fish in a school sheds a thrust type vortex street which settles down quickly to form a reverse von Kármán vortex street with constant spacing between the vortex rings.

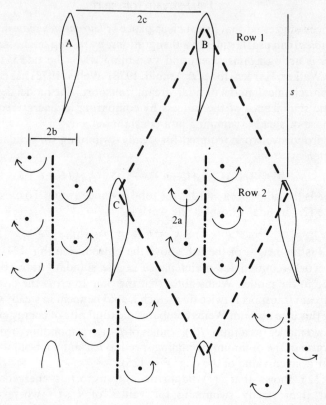

Fig. 126   This figure is explained in the text. From Weihs (1974*b*).

According to Weihs, the school should be arranged in a diamond pattern (see Fig. 126). When the ratio of the distance between fish in the same row of the school to that between vortex cores in the wake is equal to two (see Fig. 126) the force due to given motions of the fish should be twice that of a single isolated fish (Weihs, 1974*b*). Fish in the second row of the formation should experience a relative velocity that is about half of the free-stream velocity, leading to a reduction by a factor of between four and six in the required force to swim at any given speed. The implications of this result are that large increases in speed and/or range (up to a factor of about six) are possible for fish in schools over isolated individuals.

Partridge & Pitcher (1979) have recently experimentally tested the predictions made by the Weihs model using small schools (20–30 fish) of saithe (*Pollacius virens*), herring (*Clupea harengus*) and cod (*Gadus morhua*). The fish were trained to swim around a large annular tank and were observed with cameras as they did so. Partridge & Pitcher conclude that: (1) the fish generate a wake of the type suggested by Weihs; (2) the fish do not form discrete layers in the school; (3) fish have no tendency to position themselves directly between fish that are swimming in front of them (i.e. the diamond pattern was not seen); and (4) no definite phase relations were detected between the tail beats of neighbouring individuals. Points 2, 3 and 4 argue against a hydrodynamic function for schools. In addition, we may note that the findings of Partridge & Pitcher (1979) on the form of the wake conflict with those of Rosen (1959, 1961) and McCutcheon (1977). Rosen describes the wake of *Brachydanio albolineatus* as consisting of a single row of vortex rings. McCutcheon points out that it might actually be advantageous for a given fish to swim in the path of the wake of another if it were able to extract energy from it.

Weihs suggests that schools may be energetically advantageous to negatively buoyant fish. It is possible to view schools of negatively buoyant fish as being analogous to birds flying in formation. The formation flight of birds has been analysed by Lissaman & Shollenberger (1970) and Higdon & Corrsin (1978). Weihs suggests that negatively buoyant fish produce wing tip vortices from their pectoral fins and that these vortices could generate an upwash effect for some of the fish in a school. It could be argued however, that fish ought to avoid the flow from fish ahead of them because of a possible downwash effect over their own fins. As far as I am aware, there is no good experimental data on the school structure of negatively buoyant fish at present.

Porpoising

Au & Weihs (1980) describe three modes of swimming in the dolphins:
1. Slow swimming close to the surface.
2. Cruising just beneath the surface with little splashing.
3. Fast 'running', where the dolphin leaps clear of the water in a series of parabolic leaps. This swimming behaviour is commonly referred to as porpoising.

Au & Weihs have produced a simple hydromechanical model of porpoising which assumes that the animal can be likened to a rigid body. The model is designed to determine if the energy cost of leaping (porpoising) is less than that for continuous swimming close to the surface. In a similar model Blake (1983c) introduces a drag augmentation factor (κ) that accounts for the additional drag generated by bodily oscillation (see chapter 5). Blake discusses the the 'crossover' speeds at which leaping becomes energetically less costly than swimming close to the surface and out of its influence in relation to predicted maximum prolonged swimming speeds.

The energy required to traverse a given distance s can be written as

$$W_s = \tfrac{1}{2}\rho S_w U^2(\kappa C_D)s \tag{239}$$

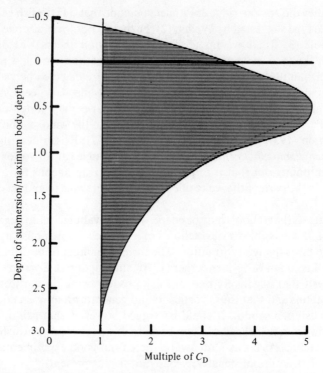

Fig. 127   Drag augmentation factor due to surface interference is plotted against a non-dimensional index of submersion depth. Based on Hertel (1966, 1969).

(Blake, 1983c, eqn 3). However, when swimming close to the surface an additional drag increment arises due to wave drag generated by the animal. For streamlined bodies the drag approximates the predicted rigid-body frictional drag when the submersion depth (h, measured from the surface to the centreline of the body) is about three times the maximum depth of the body section d. For $h/d < 3.0$ the drag is increased by a factor § (see Fig. 127) which has a maximum value of about 5 (Hertel, 1966). Drag decreases when the body emerges (i.e. $h/d > 0.5$). For fish and cetaceans a maximum value of 4.5 is assigned to § because fish are not perfect bodies of revolution (Au & Weihs, 1980). Taking account of wave drag we can write eqn 239 as

$$W_{\mathrm{s}} = \tfrac{1}{2}\rho S_{\mathrm{w}} U^2(\kappa C_{\mathrm{D}})\S s \tag{240}$$

Values of C_{D} can be calculated from empirically derived expressions such as eqn 39. For a fineness ratio of about five, (i.e. $l/d = 5$), $C_{\mathrm{D}} = 0.0024$.

The energy required for leaping is given by simply multiplying the weight of the animal (mg) by the height of its jump. The mass term in the equation must include longitudinal added mass which is about 20% of the mass of the animal. For maximum distance and height the animal leaves the water at an angle of 45° when the energy required for leaping (W_j) can be written as

$$W_j = 1.2(mg)(U^2/4g) \tag{241}$$

The energies required for subsurface swimming and leaping may be compared from the following equation

$$W_j/W_s = \frac{1.2(mg)(U^2/4g)}{\frac{1}{2}\rho S_w U^2(U^2/g)(\kappa C_D)\S} \tag{242}$$

where the term U^2/g is the maximum leaping distance. According to Bainbridge (1961), $S_w = 0.4\,l^2$ and $m = 0.01\,l^3$ for fish-like bodies of revolution. Substituting these expressions into eqn 242 and setting $k = 4$, $\S = 4.5$, $W_j/W_s = 1$ gives

$$U_{cross} = (341.0\,l)^{\frac{1}{2}} \tag{243}$$

where U_{cross} is the crossover speed at which leaping becomes energetically less expensive than subsurface swimming.

For any arguments concerning the possible energetic advantage of leaping over subsurface swimming to be valid, it must be shown that the animal is capable of attaining the required crossover speeds. Blake (1983c) equates the power required for subsurface swimming with that available from the swimming muscles

$$\tfrac{1}{2}\rho S_w U^3(\kappa C_D)\S = 0.75P_f(M_w/2) \tag{244}$$

where a propulsive (Froude) efficiency of 75% is assumed and P_f and M_w refer to the power factor of the muscles and their weight, respectively. This approach is essentially that employed in chapter 3 where swimming performance is discussed. Manipulation of eqn 244 and substitution of numerical values ($k = 4$, $\S = 4.5$, $S_w = 0.4\,l^2$) gives

$$1.33(\rho 0.4l^2 U^3 8.5C_f) = P_f M_w \tag{245}$$

which reduces further to

$$4.52\rho l^2 U^3 C_f = P_f M_w \tag{246}$$

Substituting in the expressions for C_f for the case of a laminar and turbulent boundary layer, respectively, (eqns 64 and 65) and rearranging gives

$$U_{lam} = \left(\frac{P_f M_w}{0.60\,l^{3/2}}\right)^{2/5} \tag{247}$$

$$U_{turb} = \left(\frac{P_f M_w}{0.133\,l^{9/5}}\right)^{5/14} \tag{248}$$

Assuming that 50% of the mass of the animal is propulsive musculature of which 20% is active red muscle and a power factor of 250 W kg^{-1} ($= 2.5 \times 10^6$ ergs s^{-1} g^{-1}) we can write

$$U_{lam} = 28.1\,l^{0.56} \tag{249}$$

$$U_{turb} = 33.6\,l^{0.39} \tag{250}$$

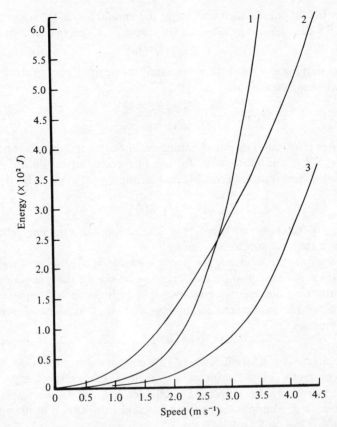

Fig. 128   The energies required for swimming close to the surface (curve 1, § = 4.5, κ = 4), leaping (curve 2) and swimming out of the influence of the surface (curve 3, § = 1, κ = 4) are plotted against swimming speed for the case of a dolphin (l = 2.15 m). From Blake (1983c).

The energies required for swimming deeply submerged (W_s, $h/d > 3.0$), close to the surface ($h/d = 0.5$) and leaping (W_j) are plotted in Fig. 128 for the case of a 'typical' dolphin (l = 2.15 m). Fig. 128 shows that: (1) leaping is energetically less efficient (i.e. energetically more costly) than swimming close to the surface up to a certain speed (U_{cross} = 2.71 m s^{-1}), after which it is more efficient; and (2) the curve for deeply submerged swimming also crosses that for leaping (U_{cross} = 5.3 m s^{-1}) and so at higher speeds leaping is the most efficient swimming strategy.

In Table 3 calculated values of U_{cross} for κ = 1 and 4 are compared with the predicted speeds for the case of a laminar boundary layer, a turbulent one and a boundary layer that is 50% laminar and 50% turbulent. Table 3 indicates that dolphins ought to be able to reach their crossover speeds even

Table 3 *Crossover speeds for the case of* $\alpha = 4$, $\S = 4.5$ (U_{cross}) *and*
$\kappa = 1.0$, $\S = 4.5$ (U'_{cross}) *are compared to the predicted maximum cruising speeds for the case of a laminar boundary layer* (U_{lam}), *a turbulent one* (U_{turb}) *and a boundary layer that is 50% laminar and 50% turbulent* ($U_{50/50}$). *From Blake (1983c)*

Length (m)	Volume (m³)	Reynolds Number (× 10⁶)	U_{cross} (m s⁻¹)	U'_{cross} (m s⁻¹)	U_{lam} (m s⁻¹)	U_{turb} (m s⁻¹)	$U_{50/50}$ (m s⁻¹)
0.21	0.0001	0.18	0.85	1.68	1.55	1.10	1.33
0.37	0.0005	0.41	1.12	2.23	2.12	1.37	1.75
0.47	0.001	0.60	1.27	2.52	2.43	1.51	1.97
0.80	0.005	1.32	1.65	3.29	3.27	1.86	2.57
1.00	0.01	1.85	1.85	3.67	3.70	2.02	2.86
1.70	0.05	4.10	2.41	4.79	4.99	2.49	3.74
2.15	0.1	5.83	2.71	5.39	5.69	2.73	4.21
2.44	0.15	7.05	2.89	5.74	6.10	2.87	4.49
2.69	0.2	8.15	3.03	6.03	6.45	2.98	4.71
2.89	0.25	9.07	3.14	6.25	6.71	3.06	4.89
3.64	0.5	12.81	3.52	7.01	7.64	3.35	5.50
4.57	1.0	18.76	3.95	7.86	8.67	3.66	6.17
7.77	5.0	40.02	5.15	10.24	11.68	4.50	8.09
9.77	10.0	56.37	5.77	11.49	13.28	4.93	9.11

in the presence of a turbulent boundary layer. It is likely that at least 50% of the body surface of these animals is subject to laminar boundary layer flow at typical cruising speeds (see chapter 4).

Larger cetaceans such as killer whales (*Orcinus orca*) also porpoise. Although the energy required for porpoising in these animals is high, it is less than that for swimming close to the surface at the same average speed. However, the baleen whales (Balaenidae, Balaenopteridae) do not porpoise. It is likely that values of U_{cross} based on $\kappa = 1$ are appropriate here, as bodily undulation is restricted largely to the flukes and caudal peduncle. For the case of an 18 m long sei whale (*Balaenoptera borealis*) the value of U_{cross} would be 15.3 m s⁻¹. This value is about three times the observed cruising speeds of sei whales of this size (Coffey, 1977). Blake (1983c) suggests that the baleen whales cruise at speeds at which their filtration apparatus (baleen) functions most effectively.

Fish do not porpoise, although certain species live close to the surface and escape from predators by leaving the water (e.g. Exocoetidae, Hemirhamphidae, Pantodontidae, Gasteropeloidae). Prior to leaving the water the Exocoetidae and Pantodontidae accelerate rapidly (burst swimming) and once airborne employ specialized aerodynamic surfaces (pectoral fins) to

(a)

(b)

(c)

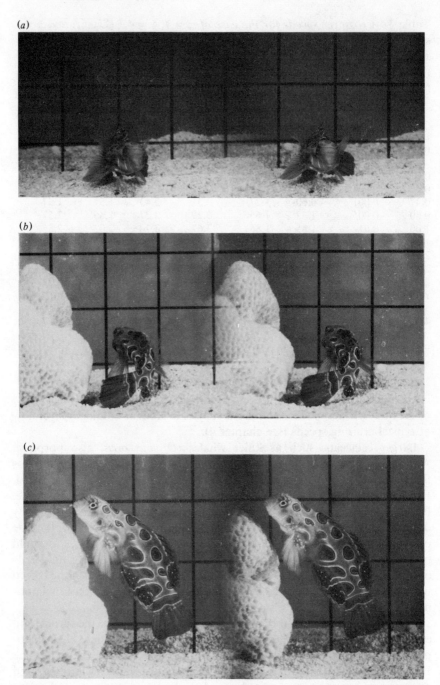

Fig. 129   A specimen of *Synchropus* hovering in (*a* and *b*) and out of (*c*) the influence of the ground. From Blake (1979*d*).

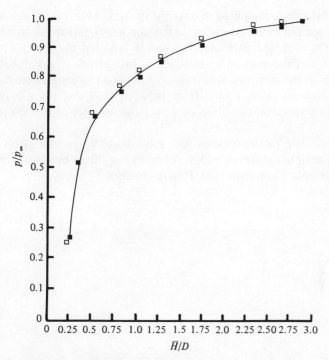

Fig. 130   The power required to hover at various heights above the ground relative to that out of the ground effect for a specimen of *Synchropus* is plotted against a non-dimensional index of height for the case of a laminar (□) and turbulent boundary layer (■) over the pectoral fins. From Blake (1979*d*).

maximize the distance travelled. This process is distinct from porpoising which can be viewed as a strategy for rapid prolonged cruising that involves no special aerodynamic surfaces. The leaping behaviour of the salmonids is also not to be confused with porpoising. Leaping in salmon usually involves a 'standing jump' which requires a high level of burst-swimming performance (Gray, 1968).

Ground effect

Blake (1979*d*) has investigated the influence of the ground effect on the energetics of hovering in the mandarin fish (*Synchropus picturatus*), a negatively buoyant coral reef fish. Fig. 129 shows a specimen of *Synchropus* hovering in the influence of the ground (Fig. 129*a* and *b*) and out of ground effect (Fig. 129*c*). Thrust and power for the pectoral fins of *Synchropus* hovering at various heights above the substrate (as indicated by the ratio of the height above the bottom J to the span of the pectoral fins D) were calculated from flow velocity data.

Values of the power required to hover at various values of J/D against the power required out of the ground effect (i.e. for $J/D > 3.0$) are plotted in Fig. 130. Fig. 130 clearly shows that the power required to hover decreases markedly as the fish comes into ground effect. In captivity at least *Synchropus* hovers close to the substrate whilst feeding on small interstitial crustaceans and algal growths. Values of J/D of between 0.25 and 0.5 are typical, corresponding to savings on induced power (see chapter 6) of the order of 30–60%.

It is likely that further studies will show that the ground effect is an important factor in the energy budgets of many negatively buoyant, demersal fish (e.g. Selachii, Callionymoidei, Heterosomata).

9

Other issues

Introduction

In this chapter aspects of fish locomotion that do not fall naturally into any of the categories that we have considered so far are discussed. Essentially we are concerned with three topics: larval swimming, locomotion in the air and on land, and rheotactic behaviour. These topics involve some concepts and ideas that are different from those discussed in the preceding chapters.

Larval fish are small and swim slowly, therefore Reynolds Numbers are low relative to values typical of adult fish. Flow at the Reynolds Numbers characteristic of larval fish (10–200) is discussed with particular emphasis on the values of force coefficients. The kinematics of larval swimming is described for steady and burst swimming. Similarities and differences to the corresponding swimming activities in adult fish are noted. A hydromechanical model of larval swimming (due to Vlymen, 1974) is outlined and employed to calculate the work required to generate both resistive and reactive forces in swimming. A different model (developed by Weihs, 1980) is employed as a basis for a discussion of changes in swimming style during development.

After briefly surveying the diversity of flying fish, a detailed description of gliding behaviour in the Exocoetidae (the most common and well-studied group of flying fish) is given. A simple mathematical model is developed that describes the horizontal and vertical forces acting on the fish. The model is employed to predict gliding distances corresponding to given emergence angles and speeds for fish of a given length.

The hydrodynamics of rheotactic behaviour in negatively buoyant demersal fish such as plaice and sole is discussed. These fish exhibit a distinct behavioural sequence associated with increasing current velocity. A simple hydromechanical model is employed to describe the conditions under which horizontal and vertical displacement can be expected to occur. The predictions of the model are compared with experimental results and discussed in relation to the observed behaviour of the fish.

Finally, terrestrial locomotion in the mudskipper (*Periophthalmus*) is discussed. The kinematics of walking in *Periophthalmus* is discussed in

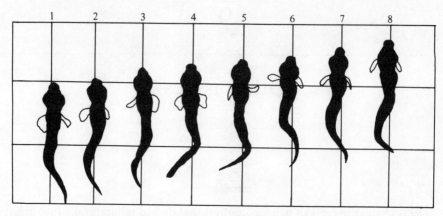

Fig. 131   Successive outlines of a larval plaice in cruising. The distance between grid lines is 5 mm and the time interval between the images is 0.028 s. Based on Batty (1981).

relation to some of the morphological modifications of the pectoral fins in this form.

Larval swimming

To this point our discussion of fish swimming mechanics has been concerned with situations where inertial effects predominate (i.e. high Reynolds Numbers). However, larval fish are small (of the order of 0.5–1.0 cm) and swim slowly. Typical Reynolds Numbers for larval fish are of the order of about 10–100. For Reynolds Numbers greater than about 200 the quadratic resistance law applies and drag is proportional to U^2. For $R_1 < 200$, two flow regimes may be distinguished corresponding to $R_1 < 10$ and $10 < R_1 < 200$, respectively. We have seen that for higher Reynolds Number flows, resistance coefficient may be calculated from empirically derived expressions (see chapter 1). Similarly, expressions are available for Reynolds Numbers between 1 and 10 (Hoerner, 1965). However, the Reynolds Number interval of 10 to 200 is a transitional zone where no such expressions can be properly applied. This makes the hydromechanical analysis of larval swimming difficult.

In contrast to the well developed literature on the propulsion of adult fish, little work has been done on swimming in larval forms. Most accounts concentrate on clupeiod larvae such as anchovy (e.g. Hunter, 1972; Vlymen, 1974) and herring (e.g. Rosenthal, 1968; Rosenthal & Hempel, 1969). Recently, Batty (1981) has given a detailed account of swimming kinematics in plaice (*Pleuronectes platessa*) larvae. Plaice larvae hatch at about 0.5 cm and metamorphose at about 12 cm after which they become benthic and lie on their left side.

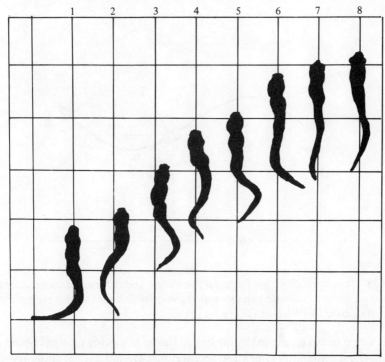

Fig. 132 Successive outlines of a larval plaice during burst swimming. The distance between grid lines is 3 mm and the time interval between frames is 0.028 s. Based on Batty (1981).

Batty describes both cruising and burst swimming (see Figs 131 and 132, respectively). In cruising it was noted that:
1. The amplitude of the propulsive body wave varies along the length of the body in a similar fashion to the adult. However, specific maximum amplitudes (amplitude divided by body length) are much greater in larval fish where the angle between the centre-line of the fish and its axis of progression may be as high as 90° at higher cruising speeds.
2. Values of U/V (ratio of forward speed to wave speed) range from about 0.2 to 0.4 and are positively correlated with U.
3. Pectoral fin movements are employed together with the body wave. The left and right pectoral fin beats are 180° out of phase.
   During burst swimming (see Fig. 132) speeds of up to 20 l s^{-1} were recorded corresponding to tail-beat frequencies of about 35 Hz. Tail-beat amplitudes are greater than those seen during cruising and the wavelength of the body wave was usually greater than the body length. No pectoral fin movements were observed during burst swimming. Batty (1981) suggests that in steady swimming the movements of the pectoral fins act to reduce the magnitude

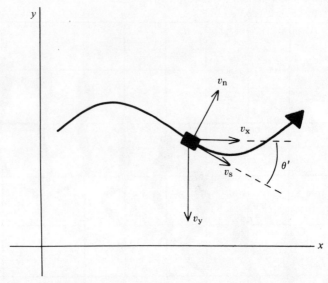

Fig. 133   Schematic diagram of a larval fish during undulatory swimming. A single element on the body is shown together with the velocity components that are discussed in the text. Based on Vlymen (1974).

of the recoil forces generated by the body. The large yawing motions observed during burst swimming (when the pectoral fins are not active) supports this view.

Vlymen (1974) has produced an interesting hydromechanical model of larval swimming which combines resistive and reactive terms. The model is applied to larval anchovy (*Engraulis mordax*) and is designed to allow calculation of the energy required to swim. Vlymen employs a model that is based on Gray & Hancock's (1955) analysis of spermatozoa propulsion. The anchovy is considered to be a ribbon of specified length attached to an inert head (see Fig. 133). Fig. 133 depicts an 'element' of the body and shows various velocity components that are generated in swimming.

The normal and spanwise resistive forces acting on an element of the body can be written as

$$dF_n = \tfrac{1}{2}\rho v_n^2 A C_n \tag{251}$$

$$dF_s = \tfrac{1}{2}\rho v_s^2 A C_s \tag{252}$$

where A is the cross-sectional area and C_n and C_s are normal and spanwise force coefficients, respectively. For low Reynolds Numbers, Hoerner (1965) and Schlichting (1952) give

$$C_n = 20.37/R_l \tag{253}$$

$$C_s = 1.33/R_l \tag{254}$$

Strictly speaking eqns 253 and 254 are only appropriate for $R_l = 1$–10, however they hold good as approximations up to R_l of about 30–40.

Multiplying each force by the element of distance (ds) in the relevant direction we can write the overall resistive work as

$$dW_{\text{resistive}} = dF_n\,ds_n + dF_s\,ds_s \tag{255}$$

The work performed in moving the head is given as the sum of a resistive and an inertial component

$$dW_{\text{head}} = \tfrac{1}{2}\rho v_x{}^3 A C_h\,dt + (m+M)(dv_x/dt)\,v_x\,dt \tag{256}$$

where C_h is a drag coefficient for the head, and m and M refer to its mass and added mass, respectively.

Next, the inertial work required to accelerate the mass and added mass of the body must be considered. These terms are more complicated than those given above, largely because they involve some awkward trigonometric terms. The virtual mass of an element is given by

$$dM = \rho\pi h^2(i)\,di \tag{257}$$

Multiplying by the acceleration gives the virtual mass force (dF_a)

$$dF_a = dMa = dM\left[\left(\frac{dv_y}{dt}\right)^2 + \left(\frac{dv_x}{dt}\right)^2\right]^{\frac{1}{2}} \tag{258}$$

The work is
$$dW_a = dF_a\,dx\cos(F_a\,dx) \tag{259}$$

Now, $dx = (v_y{}^2 + v_x{}^2)^{\frac{1}{2}}$ and $\cos(F_a\,dx) = \cos[\tan^{-1}(v_y'/v_x') - \tan^{-1}(v_y/v_x)]$, where v_y' and $v_x' = dv_y/dt$ and dv_x/dt, respectively. Finally, we have

$$dW_a = \rho\pi h^2(i)(v_y^2 + v_x^2)^{\frac{1}{2}}(v_x'^2)^{\frac{1}{2}}\sin\left(\tan^{-1}\frac{dy}{dx} - \tan^{-1}\frac{v_y'}{v_x'}\right)$$
$$\times\cos\left(\tan^{-1}\frac{v_y'}{v_x} - \tan^{-1}\frac{v_y}{v_x}\right)di\,dt \tag{260}$$

The work required to move the mass of the body can be written as

$$dW_m = dm(v_y{}^2 + v_x{}^2)^{\frac{1}{2}}(v_y'^2 + v_x'^2)^{\frac{1}{2}}\cos\left(\tan^{-1}\frac{v_y'}{v_x'} - \tan^{-1}\frac{v_y}{v_x}\right)di\,dt \tag{261}$$

The total work required is therefore

$$dW_{\text{tot}} = dW_{\text{resistive}} + dW_{\text{head}} + (dW_a + dW_m) \tag{262}$$

Eqn 262 shows that the total work required for swimming is the sum of the work required to generate the resistive forces acting on the head and body, plus that required to overcome the inertia of the head, plus the work performed in moving the mass of the body and its associated added mass.

Vlymen (1974) applied the model to a 1.4 cm specimen of *E. mordax*. At typical swimming speeds, Vlymen calculates that the resistive forces generated account for only about 8% of the total energy requirement. The inertial energy required to move the head is estimated to be only about 0.2% of the total. No mention is made of whether or not the pectoral fins are active.

Unfortunately, the framing rates employed by Batty (1981) during his study of burst swimming in plaice larvae are not high enough to allow for the calculation of dW_{head}. About 80% of the energy performed in swimming in the larval anchovy is ascribed to the inertial terms associated with the body. It would seem then, that even at relatively low Reynolds Numbers (1–100) inertial (reactive) forces are highly significant in undulatory propulsion.

Weihs (1980) has investigated the energetic significance of changes in swimming modes during growth in larval anchovy. In a hydromechanical model similar to that developed by Weihs (1974b) for burst-and-glide swimming the energy expenditures in intermittent and continuous swimming are compared. It is shown that for larvae less than about 5 mm in length continuous swimming is the more efficient mode of propulsion. However, for fish > 5 mm in length, burst-and-glide swimming becomes the more efficient mode. When the fish is about 1.5 cm in length the energy gains obtained in burst-and-glide swimming are of the same order as those predicted for adult fish. Weihs points out that small larvae (< 5 mm) swim at low R_l where viscous forces are dominant and that therefore coasting is not possible.

Flying fish

Passive (gliding) flight occurs among three teleost families, the Exocoetidae, Pantodontidae and Hemirhamphidae. Reports of gliding behaviour in the gurnards (Dactylopteridae) and barbels (e.g. *Esomus*) are thought to be mistaken (Aleyev, 1977). The Gasteropeleoidae (e.g. *Gasteropelecus*) are reported to be active (flapping) fliers. Much of the descriptive literature on flying fishes is reviewed by Aleyev (1977). Most work has concentrated on the Exocoetidae of which there are about 40 species. The exocoetids are characterized by large, membranous, wing-like pectoral fins. Some species possess two wings (e.g. *Exocoetus volitans*) and others have four (e.g. *Cypselurus heterurus*). The largest exocoetids are of the order of 50 cm in length.

Hertel (1966) summarizes the reports of the gliding behaviour of the exocoetids. The fish are reported to accelerate rapidly beneath the surface up to velocities of about 10 m s⁻¹ after which they are thought to emerge at an angle of about 30° relative to the surface of the water. The ventral lobe of the caudal fin remains in the water beating at frequencies as high as 35 Hz as the fish 'taxies' across the surface towards its 'take-off' speed which may be as high as 20 m s⁻¹. After a glide of about 50 m or so the fish falls back into the water and the process is repeated.

The equilibrium and swimming performance of the exocoetids is also discussed by Hertel (1966). The weight of the body is supported by a vertical force component from the caudal fin and the lift force that is generated by the pectoral fins. The horizontal component of thrust produced by the caudal

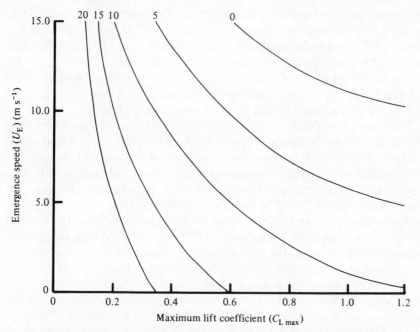

Fig. 134 Emergence speed is plotted against C_{Lmax} for various values of U_{w} (headwind speed) for a 35 cm specimen of *Cypselurus*. Based on Hertel (1966).

fin must overcome the drag force acting on the body and accelerate the mass of the fish. The balance of vertical forces acting on the animal as it taxies may be written as

$$T\sin\theta_{\text{e}} + \tfrac{1}{2}\rho U_{\text{a}}^2 A_{\text{p}} C_{\text{Lmax}} = mg \qquad (263)$$

where θ_{e} is the emergence angle, A_{p} is the area of the pectoral fins and T is the thrust force generated by the caudal fin. The velocity term in the equation (U_{a}) is measured relative to the air.

The minimum speed required for taxiing is given by

$$U_{\text{a(min)}} = \left(\frac{mg - T\sin\theta_{\text{e}}}{\tfrac{1}{2}\rho A_{\text{p}} C_{\text{Lmax}}}\right)^{\frac{1}{2}} - U_{\text{w}} \qquad (264)$$

where U_{w} is the velocity of any headwind present. Hertel (1966) assumes values for A_{p}, T, mg and C_{Lmax} for a 35 cm specimen of *Cypselurus* and reduces eqn 264 to

$$U_{\text{a(min)}} = \frac{11.3}{(C_{\text{Lmax}})^{\frac{1}{2}}} - U_{\text{w}} \qquad (265)$$

Emergence speed is plotted against assumed possible values of C_{Lmax} for various values of U_{w} in Fig. 134.

In the absence of a headwind (i.e. $U_{\text{w}} = 0$) a lift coefficient of about 1.3 would be required for an emergence speed of about 10 m s^{-1}. Shoulejkin

(1929) measured the aerodynamic properties of a model of *Exocoetus volitans* in a wind tunnel and reports a lift to drag ratio of 11 for the pectoral fins for $\alpha = 15°$. It seems reasonable to suppose that values of C_{Lmax} of the order of 1.2–1.3 are possible for the pectoral fins of flying fishes. With $C_{Lmax} = 1.2$ and a headwind of 5 m s^{-1} the emergence speed can be reduced to about 5.5 m s^{-1}. However, even this speed still requires a very high level of swimming performance (about 15 l s^{-1}) prior to emergence.

Emergence speed may be further reduced if: (1) values of T are higher than the values arbitrarily supposed by Hertel; and (2) emergence occurs at angles greater than 30°. An increase in the propulsive force would increase the contribution to the vertical force balance from the caudal fin. If the fish were to emerge at an angle greater than 30° the force required from the caudal fin for vertical support would be further reduced. With less force required for vertical equilibrium more force would be available to increase the forward speed of the fish. The contribution of the pectoral fins to the vertical equilibrium of the fish will increase rapidly as forward speed increases, making an increase in θ_e to the take-off value of 30° possible. Once clear of the water the resistance on the fish is negligible making high take-off speeds possible. More detailed observations on the emergence and taxiing behaviour of the exocoetids are needed before their performance can be properly evaluated.

The Gasteropeleoidae are reported to be capable of active flapping flight. Most Gasteropeleoidae are less than 10 cm in length. Fish of this size are said to be able to fly for distances of up to 12 m before falling back into the water. It is thought by some that the keel-like anterior end of the body remains in the water and that the caudal fin may contribute to propulsion.

Hydrodynamics of rheotaxis

It has been noted that plaice (*Pleuronectes platessa*) in a flume respond to a water current with a clearly defined pattern of rheotactic behaviour (Arnold, 1969). For fish greater than about 10 cm in length the following is observed:
1. In still water all of the fins are furled and the undersurface of the fish is in firm contact with the bottom.
2. At low current speeds (1–2 cm s^{-1}) the head is orientated in the upstream direction. The marginal fins are closely apposed to the bottom ('clamped down posture').
3. At high current speeds (20–30 cm s^{-1}) an 'arched-back posture' appears and active beating of the marginal fins is seen. As current speeds are increased further, the fish begins to be displaced downstream.

Arnold & Weihs (1978) employ a simple hydromechanical model in an investigation of rheotactic behaviour in plaice. The fish can be likened to a semi-ellipsoid with its blunt end directed upstream. The plaice is characterized

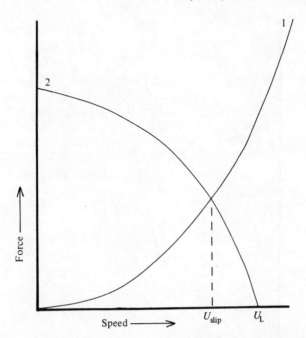

Fig. 135   This figure is explained in the text. Based on Arnold & Weihs (1978).

by a fineness ratio of about 14. For asymmetric ellipsoids attached to a plane wall the optimum fineness ratio for minimum drag is between two to three times that for a streamlined body in free flow (i.e. l/d should lie between 10 and 15 for minimum drag; Hoerner, 1965). It seems reasonable to assume that *Pleuronectes* orientates itself parallel to the flow because it experiences the least drag in this position. The drag force acting on the body of the fish is resisted by the frictional force acting between it and the substrate. When viscous drag exceeds the frictional force between the fish and the bottom the animal is displaced downstream.

However, the body of the plaice also produces a lift force. When this force exceeds the submerged weight of the fish it will be lifted off the bottom. The speed at which 'lift-off' occurs (U_L) can be expected to be greater than that at which the fish begins to slip downstream (U_{slip}). The forces acting on the fish in relation to current speed are indicated in Fig. 135. Until the drag acting on the fish exceeds the frictional force due to its submerged weight, no thrust forces will have to be generated. Similarly, no work will need to be done in producing any downwardly directed forces until the lift force acting on the body exceeds the submerged weight.

Slipping speed can be predicted by equating the horizontal forces acting

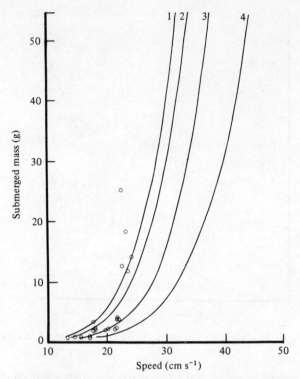

Fig. 136   Slipping speed (curve 1) and lift-off speed for $C_L = 1.0$, 1.4 and 1.8 (curves 2, 3 and 4, respectively) are plotted against submerged weight for plaice. Open circles indicate experimental observations of U_{slip}. Based on data from Arnold & Weihs (1978).

on the fish (i.e. by equating drag and the frictional resistance to drag), we have

$$\tfrac{1}{2}\rho A U^2 C_D = c_f(mg)_w \tag{266}$$

where A is the cross-sectional area of the fish, c_f is a coefficient of static friction and $(mg)_w$ is the submerged weight. The slipping speed can be calculated from

$$U_{slip} = \left[\frac{2c_f(mg)_w}{\rho A C_D}\right]^{\tfrac{1}{2}} \tag{267}$$

Similarly, for U_L we have

$$(mg)_w - \tfrac{1}{2}\rho A U^2 C_L = 0 \tag{268}$$

and

$$U_L = \left[\frac{(mg)_w}{\tfrac{1}{2}\rho A C_{Lmax}}\right]^{\tfrac{1}{2}} \tag{269}$$

Calculated values of U_{slip} and U_L for fish of different submerged weights are shown in Fig. 136. Values of U_L are calculated for $C_L = 1.0$, 1.4 and 1.8.

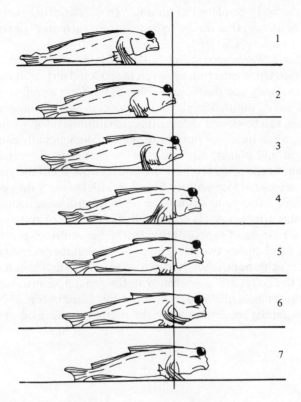

Fig. 137   Fin movements in *Periophthalmus* during locomotion on land. Successive images are at 0.06 s intervals. Based on Harris (1960).

Experimental observations made by Arnold & Weihs (1978) are also plotted in Fig. 136. Fig. 136 shows that as C_L increases the value of U_L decreases. The values of U_L and U_{slip} are equal when $C_L = 2.0$. Arnold & Weihs (1978) conclude that the observed rheotactic behaviour has evolved to maximize U_L.

Terrestrial locomotion

Some fish habitually emerge from the water and spend considerable periods of time on the land. A good example is the mudskipper, *Periophthalmus*. Harris (1960) has studied the kinematics of terrestrial locomotion in *P. koelreuteri*. Although *P. koelreuteri* possesses the same basic bony and muscular pectoral-fin elements as other teleosts there are some striking differences in their arrangement. The proximal region of the pectoral fin in *P. koelreuteri* is made up of four bones which form a rigid plate which articulates with the shoulder by means of a long vertical hinge. This allows the fin to be flattened against the side of the body or placed at an angle to it. The distal part of the fin contains 14 bony rays that can be separated from each other or collapsed to function as a single blade.

In the water *P. koelreuteri* swims in the subcarangiform mode much like many other fish. Upon emerging however, the pectoral fins are abducted from the sides of the body and the fin-rays collapsed to form a stiff, vertical strut. Gray (1968) points out that whilst in this position the proximal segment of the pectoral fin can be viewed as being functionally equivalent to the humerus of tetrapods, while the distal portion of the fin is functionally equivalent to the lower arm and plantar surface. In *Periophthalmus* the pelvic fins are located on about the same level on the body as the pectorals and are also employed in terrestrial locomotion. Together with the body the pectoral and pelvic fins form a five-point suspension when the fish is stationary. During locomotion this arrangement is only stable if the centre of gravity of the fish falls within a triangle of support provided by the centre of pressure of the tail on the ground and by two fins. Harris noted that the pectoral and pelvic fins operate synchronously but 180° out of phase, so that two fins (one on each side of the body) are in contact with the ground at any given time. A cycle of fin movements in *Periophthalmus* is illustrated in Fig. 137. It can be seen that the animal accelerates when the pectoral fins are on the ground.

References

Affleck, R. J. (1950). Some points in the function, development and evolution of the tail in fishes. *Proceedings of the Zoological Society of London*, **120**, 349–68.

Alexander, R. McN. (1965) The lift produced by the heterocercal tails of selachii. *Journal of Experimental Biology*, **43**, 131–38.

Alexander, R. McN. (1967) *Functional design in fishes*. London: Hutchinson.

Alexander, R. McN. (1969) The orientation of muscle fibres in the myomeres of fishes. *Journal of the Marine Biological Association of the United Kingdom*, **49**, 263–90.

Alexander, R. McN. (1973) Muscle performance in locomotion and other strenuous activities. In *Comparative physiology*, eds L. Bolis, K. Schmidt-Nielsen & S. H. P. Maddrell, pp. 1–21. Amsterdam: North Holland.

Alexander, R. McN. (1977) Swimming. In *Mechanics and energetics of animal locomotion*, eds R. McN. Alexander & G. Goldspink, pp. 222–48. London: Chapman & Hall.

Alexander, R. McN. (1983). The history of fish mechanics. In *Fish biomechanics*, eds P. W. Webb & D. Weihs. New York: Praeger Press. (In press)

Aleyev, Yu G. (1963) The functional principles of external structure of fishes. USSR Academy of Sciences Press, M.1–247.

Aleyev, Yu. G. (1977) *Nekton*. The Hague: Junk.

Allen, W. H. (1961) Underwater flow visualization techniques. *Technical Publications of the U.S. Navy Ordinance Test Station*, no. 2759, 1–28.

Arnold, G. P. (1969) The reactions of the plaice (*Pleuronectes platessa* L.) to water currents. *Journal of Experimental Biology*, **51**, 681–97.

Arnold, G. P. & Weihs, D. (1978) The hydrodynamics of rheotaxis in the plaice (*Pleuronectes plattessa* L.). *Journal of Experimental Biology*, **75**, 147–69.

Aristotle (fourth century B.C.) *The movement of animals*. English translation by E. S. Forster (1931). London: Heinemann.

Au, D. & Weihs, D. (1980) At high speeds dolphins save energy by leaping. *Nature, London*, **284**, 348–50.

Bainbridge, R. (1958) The speed of swimming of fish as related to size and to the frequency and amplitude of the tail beat. *Journal of Experimental Biology*, **35**, 109–33.

Bainbridge, R. (1960) Speed and stamina in three fish. *Journal of Experimental Biology*, **37**, 129–53.

Bainbridge, R. (1961) Problems of fish locomotion. In *Vertebrate locomotion*, Symposium of the Zoological Society of London, **5**, 13–32.

193

Bainbridge, R. (1963) Caudal fin and body movements in the propulsion of some fish. *Journal of Experimental Biology*, **40**, 23–56.

Batty, R. S. (1981) Locomotion of plaice larvae. In *Vertebrate locomotion*, Symposium of the Zoological Society of London, **48**, 53–69.

Beamish, F. W. H. (1978) Swimming capacity. In *Fish physiology*, vol. 7, *Locomotion*, eds W. S. Hoar & D. J. Randall, pp. 101–87. London: Academic Press.

Blake, R. W. (1976) On seahorse locomotion. *Journal of the Marine Biological Association of the United Kingdom*, **56**, 939–49.

Blake, R. W. (1977) On ostraciiform locomotion. *Journal of the Marine Biological Association of the United Kingdom*, **57**, 1047–55.

Blake, R. W. (1978) On balistiform locomotion. *Journal of the Marine Biological Association of the United Kingdom*, **58**, 73–80.

Blake, R. W. (1979a) *The mechanics of labriform locomotion*. Unpublished Ph.D. Thesis, University of Cambridge.

Blake, R. W. (1979b) The mechanics of labriform locomotion I. Labriform locomotion in the Angelfish (*Pterophyllum eimekei*): an analysis of the power stroke. *Journal of Experimental Biology*, **82**, 255–71.

Blake, R. W. (1979c) The swimming of Mandarin Fish *Synchropus picturatus* (Callinyiidae: Teleostei). *Journal of the Marine Biological Association of the United Kingdom*, **59**, 421–28.

Blake, R. W. (1979d) The energetics of hovering in the Mandarin Fish (*Synchropus picturatus*). *Journal of Experimental Biology*, **82**, 25–33.

Blake, R. W. (1980a) The mechanics of labriform locomotion II: An analysis of the recovery stroke and the overall fin-beat cycle propulsive efficiency in the Angelfish. *Journal of Experimental Biology*, **85**, 337–42.

Blake, R. W. (1980b) Undulatory median fin propulsion of two teleosts with different modes of life. *Canadian Journal of Zoology*, **58**, 2116–19.

Blake, R. W. (1981a) Influence of pectoral fin shape on thrust and drag in labriform locomotion. *Journal of Zoology, London*, **194**, 53–66.

Blake, R. W. (1981b) Mechanics of ostraciiform locomotion. *Canadian Journal of Zoology*, **59**, 1067–71.

Blake, R. W. (1981c) Mechanics of drag-based mechanisms of propulsion in aquatic vertebrates. In *Vertebrate locomotion*, Symposium of the Zoological Society of London, **48**, 29–52.

Blake, R. W. (1983a) Median and paired fin propulsion. In *Fish biomechanics*, eds P. W. Webb & D. Weihs. Praeger Press. (In press).

Blake, R. W. (1983b) Swimming in the electric-eels and knifefishes. *Canadian Journal of Zoology*. (In press)

Blake, R. W. (1983c) Energetics of leaping in dolphins and other aquatic animals. *Journal of the Marine Biological Association of the United Kingdom*. (In press)

Blaxter, J. H. S. (1969) Swimming speeds of fish. *Food and Agriculture Organisation, Fisheries Report*, **62** (2), 69–100.

Boddeke, R. E., Slijper, E. J. & van der Stelt, A. (1959) Histological characteristics of the body musculature of fishes in connection with their mode of life. *Proceedings, Koninklijke (Nederlandse) Akademie van Wetenschappen (Series C)*, **62**, 589–93.

Bone, Q. (1966) On the function of the two types of myotomal muscle fibres in elasmobranch fish. *Journal of the Marine Biological Association of the United Kingdom*, **46**, 321–49.

Bone, Q. (1971) On the scabbard fish, *Aphanpus carbo*. *Journal of the Marine Biological Association of the United Kingdom*, **51**, 219–25.

Bone, Q. (1972) Buoyancy and hydrodynamic functions of the integument in the castor-oil fish, *Ruvettus pretiogus* (Pisces: Gempylidae). *Copeia*, **78**, 78–87.

Bone, Q. (1974) Muscular and energetic aspects of fish swimming. In *Swimming and flying in nature*, vol. 2, eds T. Y. Wu, C. J. Brokaw & C. Brennen, pp. 493–528. New York: Plenum.

Bone, Q. (1978) Locomotor muscle. In *Fish physiology*, vol. 7, eds W. S. Hoar & D. J. Randall, pp. 361–424. London: Academic Press.

Bone, Q. & Brook, C. E. R. (1973) On the *Schedophilus medusophagus* (Pisces: Stromateoidei). *Journal of the Marine Biological Association of the United Kingdom*, **53**, 753–61.

Bone, Q. & Howarth, J. V. (1967) *Council Report, Marine Biological Association of the United Kingdom for* 1966–67.

Borelli, G. (1680) *De motu animalium*, pars I. Bernabo: Rome.

Bramwell, A. R. S. (1976) *Helicopter dynamics*. London: Edward Arnold.

Breder, C. M. (1926) The locomotion of fishes. *Zoologica*, **4**, 159–256.

Breder, C. M. (1967) On the survival value of fish schools. *Zoologica*, **52**, 25–40.

Brett, J. R. (1963) The energy required for swimming by young sockeye salmon with a comparison of the drag force on a dead fish. *Transactions of the Royal Society of Canada*, **1**, 441–57.

Brett, J. R. (1964) The respiratory metabolism and swimming performance of young sockeye salmon. *Jounal of the Fisheries Research Board of Canada*, **21**, 1183–1226.

Brett, J. R. (1967) Swimming performance of sockeye salmon in relation to fatigue time and temperature. *Journal of the Fisheries Research Board of Canada*, **24**, 1731–41.

Brett, J. R. & Sutherland, D. B. (1965) Respiratory metabolism of pumpkin-seed, *Lepomis gibbosus*, in relation to swimming speed. *Journal of the Fisheries Research Board of Canada*, **22**, 405–9.

Brock, V. E. & Riffenburgh, R. H. (1960) Fish schooling: a possible factor in reducing predation. *Journal du Conseil*, **25**, 307–17.

Burdak, V. D. (1969) The ontogenetic development of the scale cover of the mullet *Mugil saliens* Risso. *Zoological Journal*, **47**, 732–38.

Carey, F. G. & Teal, J. M. (1969) Regulation of body temperature by bluefin tuna. *Comparative Biochemistry and Physiology*, **28**, 205–13.

Chopra, M. G. (1974*a*) Hydromechanics of lunate-tail swimming propulsion. *Journal of Fluid Mechanics*, **64**, 375–91.

Chopra, M. G. (1974*b*) Lunate-tail swimming propulsion. In *Swimming and flying in nature*, vol. 2, eds T. Y. Wu, C. J. Brokaw & C. Brennen, pp. 635–52. New York: Plenum.

Coffey, D. J. (1977) *Sea mammals*. London: Hart-Davis, MacGibbon.

Daugherty, R. L. & Franzini, J. B. (1977) *Fluid mechanics with engineering applications*. New York: McGraw Hill.

Denton, E. J. (1961) The buoyancy of fish and cephalopods. *Progress in Biophysics*, **2**, 179–236.

Dickinson, S. (1928) The dynamics of bicycle pedalling. *Proceedings of the Royal Society of London (B)*, **103**, 225–33.

Dorn, P., Johnson, L. & Darby, C. (1979) The swimming performance of nine species of common California in-shore fishes. *Transactions of the American Fisheries Society*, **108**, 366–72.

Driedzic, W. R. & Hochachka, P. W. (1978) Metabolism in fish during exercise. In *Fish physiology*, vol. 7, eds W. S. Hoar & D. J. Randall, pp. 503–43. London: Academic Press.

Du Bois, A. B. & Ogilvy, C. S. (1978) Forces on the tail surface of swimming fish: thrust, drag and acceleration in bluefish (*Pomatomus saltatrix*). *Journal of Experimental Biology*, **77**, 225–41.

Ellington, C. P. (1978) The aerodynamics of normal hovering flight: three approaches. In *Comparative physiology: water, ions and fluid mechanics*, ed. L. Bolis, K. Schmidt-Nielsen & S. H. P. Maddrell, pp. 327–45. Cambridge University Press.

Fierstine, H. L. & Walters, V. (1968) Studies of locomotion and anatomy of scombriod fishes. *Memoir of the Southern California Academy of Sciences*, **6**, 1–31.

Flitney, F. W. & Johnston, I. A. (1979) Mechanical properties of isolated fish red and white muscle fibres. *Proceedings of the Physiological Society*, **295**, 49–50.

Freadman, M. A. (1979) Swimming energetics of striped bass (*Morone saxatilis*) and bluefish (*Pomatomus saltatrix*): gill ventilation and swimming metabolism. *Journal of Experimental Biology*, **83**, 217–30.

Freadman, M. A. (1981) Swimming energetics of striped bass (*Morone saxatilis*) and bluefish (*Pomatomus saltatrix*): hydrodynamic correlates of locomotion and gill ventilation. *Journal of Experimental Biology*. (In press)

Gadd, G. E. (1964) Some hydrodynamical aspects of swimming. *National Physical Laboratory, Ship Division Report*, number 45, 1–22.

Gero, D. R. (1952) The hydrodynamic aspects of fish propulsion. *American Museum Notivates*, **1601**, 1–32.

Goldspink, G. (1977) Muscle energetics and animal locomotion. In *Mechanics and energetics of animal locomotion*, eds R. McN. Alexander & G. Goldspink, pp. 57–81. London: Chapman & Hall.

Goldstein, S. (1938) *Modern developments in fluid dynamics*, vol. 2. Oxford: Clarendon Press.

Gray, J. (1933a) Studies in animal locomotion I. The movement of fish with special reference to the eel. *Journal of Experimental Biology*, **10**, 88–104.

Gray, J. (1933b) Studies in animal locomotion II. The relationship between waves of muscular contraction and the propulsive mechanism of the eel. *Journal of Experimental Biology*, **10**, 386–90.

Gray, J. (1933c) Studies in animal locomotion III. The propulsive mechanism of the whiting. *Journal of Experimental Biology*, **10**, 391–400.

Gray, J. (1936) Studies in animal locomotion VI. The propulsive powers of the dolphin. *Journal of Experimental Biology*, **13**, 192–99.

Gray, J. (1957) *How animals move*. London: Penguin Books.

Gray, J. (1968) *Animal locomotion*. London: Weidenfeld & Nicolson.

Gray, J. & Hancock, G. J. (1955) The propulsion of sea-urchin spermatozoa. *Journal of Experimental Biology*, **32**, 802–14.

Greenwalt, C. H. (1975) The flight of birds. *Transactions of the American Philosophical Society*, **65**, 1–67.

Greer-Walker, M., Harden-Jones, F. R. & Arnold, G. P. (1978) Movements of plaice (*Pleuronectes platessa*) tracked in the open sea. *Journal du Conseil*, **38**, 58–86.

Greer-Walker, M. & Pull, G. (1975) A survey of red and white muscle in marine fish. *Fishery Biology*, **7**, 294–300.

Gregory, W. K. (1928) Studies on the body forms of fishes. *Zoologica*, **8**, 325–421.

Grenholm, A. (1923) *Studien über die flossemmuskulatur der Teleostier.* Unpublished Ph.D. Thesis, University of Uppsala.

Grove, A. J. & Newell, G. E. (1936) A mechanical investigation into an effectual action of the caudal fin of some aquatic chordates. *Annual Magazine of Natural History*, **17**, 280–90.

Hansen, R. J. (1973) The reduced drag of polymer solutions in turbulent and transient laminar shear flows. *Transactions of the American Society of Mechanical Engineering, Journal of Fluids Engineering*, **1**, 1–23.

Harris, J. E. (1935) The swimming movements of fishes. *Annual Report of the Tortugas Laboratory, Carnegie Institute*, 1934–35, 251–53.

Harris, J. E. (1936) The role of the fins in the equilibrium of the swimming fish I. Wind tunnel test on a model of *Mustelus canis* (Mitchell). *Journal of Experimental Biology*, **13**, 476–93.

Harris, J. E. (1937*a*) The mechanical significance of the position and movements of the paired fins in the Teleostei. *Papers of the Tortugas Laboratory, Carnegie Institute*, **31**, 173–89.

Harris, J. E. (1937*b*) The role of fin movements in the equilibrium of the fish. *Annual Reports of the Tortugas Laboratory, Carnegie Institute*, 1936–37, 91–3.

Harris, J. E. (1938) The role of the fins in the equilibrium of the swimming fish II. The role of the pelvic fins. *Journal of Experimental Biology*, **15**, 32–47.

Harris, J. E. (1950) *Diademodus hydei*, a new fossil shark from Cleveland shale. *Proceedings of the Zoological Society of London*, **120**, 683–97.

Harris, J. E. (1953) Fin patterns and mode of life in fishes. In *Essays in marine biology*, eds S. M. Marshall & P. Orr, pp. 17–28. Edinburgh and London: Oliver & Boyd.

Harris, V. A. (1960) On the locomotion of the mud-skipper *Periophthalmus koelreuteri*, (Pallas) (Gobiidae). *Proceedings of the Zoological Society of London*, **134**, 107–35.

Henderson, Y. & Haggard, H. W. (1925) The maximum of human power and its fuel. *American Journal of Physiology*, **72**, 264–82.

Hertel, H. (1966) *Structure, form and movement.* New York: Reinhold.

Hertel, H. (1969) Hydrodynamics of swimming and wave-riding dolphins. In *The biology of marine mammals*, ed. H. T. Anderson, pp. 31–63. London: Academic Press.

Higdon, J. J. L. & Corrsin, S. (1978) Induced drag of a bird flock. *American Naturalist*, **112**, 727–44.

Hill, A. V. (1939) The transformation of energy and mechanical work of muscles. *Proceedings of the Physical Society*, **51**, 1–18.

Hill, A. V. (1950) The dimensions of animals and their muscular dynamics. *Science Progress*, **38**, 209–30.

Hill, A. V. (1964) The efficiency of mechanical power development during muscular shortening and its relation to load. *Proceedings of the Royal Society* (*B*), **159**, 319–24.

Hobson, E. S. (1974) Feeding relationships of the teleostean fishes on coral reefs in Kona, Hawaii. *Fisheries Bulletin of the United States*, **72**, 914–1031.

Hoerner, S. F. (1965) *Fluid dynamic drag*. Published by the author.

Hora, S. L. (1935) Ancient Hindu conceptions of correlation between form and locomotion of fishes. *Journal of Asiatic Society*, **1**, 1–7.

Houssay, S. F. (1912) *Forme, prissance et stabilite des poissons*. Paris: Hermann.

Hoyt, J. W. (1974) Hydrodynamic drag reduction due to fish slimes. In *Swimming and flying in nature*, eds T. Y. Wu, C. J. Brokaw & C. Brennen, pp. 653–72. New York: Plenum.

Hunter, J. R. (1971) Sustained speed of jack mackerel, *Trachurus symmetricus*. *Fisheries Bulletin of the United States*, **69**, 267–71.

Hunter, J. R. (1972) Swimming and feeding behaviour of larval anchovy, *Engraulis mordax*. *Fisheries Bulletin of the United States*, **70**, 821–38.

Hunter, J. R. & Zweifel, J. R. (1971) Swimming speed, tail-beat frequency, tail-beat amplitude and size in jack mackerel, *Trachurus symmetricus*, and other fishes. *United States Fisheries and Wildlife Service, Fisheries Bulletin*, **69**, 253–66.

Jarman, G. M. (1961) A note on the shape of fish myotomes. In *Vertebrate locomotion*, Symposium of the Zoological Society of London, **5**, 33–35.

Johnston, I. A., Davison, W. & Goldspink, G. (1977) Energy metabolism of carp swimming muscle. *Journal of Comparative Physiology*, **114**, 203–16.

Johnston, I. A., Frearson, N. & Goldspink, G. (1972) Myofibrillar ATPase activities of red and white muscles of marine fish. *Experientia*, **28**, 713–14.

Johnston, I. A. & Goldspink, G. (1973) Quantitative studies on muscle glycogen utilization in Crucian carp (*Carassius carassius* L.). *Journal of Experimental Biology*, **59**, 607–15.

Johnston, I. A., Walesby, N. J., Davison, W. & Goldspink, G. (1975) Temperature adaptation in the myosin of an antarctic fish. *Nature, London*, **254**, 74–5.

Kashin, S. M. & Smoljaninov, V. V. (1969) Concerning the geometry of fish trunk muscles. *Journal of Ichthyology* (*USSR*), **9**, 923–25.

Kempf, G. & Neu, W. (1932) Schleppversuche mit hechten zur Messung des Wassersidestancles. *Zeitschrift für vergleichende Physiologie*, **17**, 353–64.

Kermack, K. A. (1948) The propulsive powers of the Blue and Fin Whales. *Journal of Experimental Biology*, **20**, 23–7.

Kohn, M. C. (1973) Energy storage in drag reduction polymer solutions. *Journal of Polymer Science*, **11**, 2339.

Kramer, M. O. (1960*a*) Boundary layer stabilization by distributed damping. *Journal of the American Society of Naval Engineering*, **72**, 25–33.

Kramer, M. O. (1960*b*) The dolphin's secret. *New Scientist*, **7**, 118–20.

Lang, T. G. (1966) Hydrodynamic analysis of cetacean performance. In *Whales, dolphins and porpoises*, ed. K. S. Norris, pp. 410–32. Berkeley: University of California Press.

Lang, T. G. (1974) Speed, power, and drag measurements of dolphins and porpoises. In *Swimming and flying in nature*, eds T. Y. Wu, C. J. Brokaw & C. Brennen, pp. 553–72. New York: Plenum Press.

Lang, T. G. & Daybell, D. A. (1963) Porpoise performance tests in a seawater tank. *Naval Ordinance Test Station Technical Report*, 3063, 1–50.

Lighthill, M. J. (1960) Note on the swimming of slender fish. *Journal of Fluid Mechanics*, **9**, 305–17.

Lighthill, M. J. (1969) Hydromechanics of aquatic animal propulsion: a survey. *Annual Review of Fluid Mechanics*, **1**, 413–46.

Lighthill, M. J. (1970) Aquatic animal propulsion of high hydromechanical efficiency. *Journal of Fluid Mechanics*, **44**, 265–301.

Lighthill, M. J. (1971) Large-amplitude elongated body theory of fish locomotion. *Proceedings of the Royal Society (B)*, **179**, 125–38.

Lighthill, M. J. (1974) Aerodynamic aspects of animal flight. In *Swimming and flying in nature*, vol. 2, eds T. Y. Wu, C. J. Brokaw & C. Brennen, pp. 423–91. New York: Plenum.

Lighthill, M. J. (1975) *Mathematical biofluiddynamics*. Philadelphia: Society for Industrial and Applied Mathematics.

Lighthill, M. J. (1977) Mathematical theories of fish swimming. In *Fisheries mathematics*, ed. J. H. Steele, pp. 131–44. New York: Academic Press.

Lindsey, C. C. (1978) Form, function and locomotory habits in fish. In *Fish physiology*, vol. 7, eds W. S. Hoar & D. J. Randall, pp. 1–100. London: Academic Press.

Lissaman, P. B. S. & Shollenberger, C. A. (1970) Formation flight of birds. *Science*, **168**, 1003–5.

Lumley, J. L. (1969) Drag reduction by additives. *Annual Review of Fluid Mechanics*, **3**, 367–84.

McCutcheon, C. W. (1970) The trout tail fin: a self-cambering hydrofoil. *Journal of Biomechanics*, **3**, 271–81.

McCutcheon, C. W. (1977) Froude propulsive efficiency of a small fish measured by wake visualization. In *Scale effects in animal locomotion*, ed. T. J. Pedley, pp. 339–63. London: Academic Press.

Magnan, A. (1930) Les caractéristiques géometriques et physiques des poissons. *Annales des Sciences naturelles*, **13**, 1971–81.

Magnan, A. & St Laque, A. (1929) Essai de théorie du poisson. *Bulletin Technique Aéronautique*, **50**, 1–180.

Magnuson, J. J. (1966) Continuous locomotion in scombriod fishes. *American Zoologist*, **6**, 5.

Magnuson, J. J. (1970) Hydrostatic equilibrium of *Euthynnus affinis*, a pelagic teleost without a gas bladder. *Copeia*, 1970, 56–85.

Magnuson, J. J. (1972) Comparative study of adaptations for continuous swimming and hydrostatic equilibrium of scombriod and xiphoid fishes. *United States Fisheries and Wildlife Service, Fisheries Bulletin*, **71**, 337–56.

Magnuson, J. J. (1978) Locomotion by scombriod fishes: hydrodynamics, morphology and behaviour. In *Fish physiology*, vol. 7, eds W. S. Hoar & D. J. Randall, pp. 239–313. London: Academic Press.

Magnuson, J. J. & Prescott, J. H. (1966) Courtship, feeding and miscellaneous behaviour of Pacific Bonito (*Sarda chiliensis*). *Animal Behaviour*, **14**, 54–67.

Marey, E. J. (1894) *Le mouvement*. Paris: Masson.

Newman, J. N. & Wu, T. Y. (1973) A generalized slender-body theory for fish-like forms. *Journal of Fluid Mechanics*, **57**, 673–93.

Newman, J. N. & Wu, T. Y. (1974) Hydromechanical aspects of fish swimming. In *Swimming and flying in nature*, vol. 2, eds T. Y. Wu, C. J. Brokaw & C. Brennen, pp. 615–34. New York: Plenum.

Nishi, S. (1938) Muskelsystem II. Muskelu des rumples. *Handbuch der vergleichenden Anatomie der Wirbeltiere*. Berlin: Urban & Schwartzenberg.

Nursall, J. R. (1956) The lateral musculature and the swimming of fish. *Proceedings of the Zoological Society of London*, **126**, 127–43.

Nursall, J. R. (1958) The caudal fin as a hydrofoil. *Evolution*, **12**, 116–20.

Ovchinnikov, V. V. (1966) Turbulence in the boundary layer as a method for reducing the resistance of certain fish to movement. *Biophysics*, **11**, 186–8.

Parry, D. A. (1949) The swimming of whales and a discussion of Gray's Paradox. *Journal of Experimental Biology*, **26**, 24–34.

Partridge, B. L. & Pitcher, J. J. (1979) Evidence against a hydrodynamic function for fish schools. *Nature, London*, **279**, 418–19.

Pedley, T. J. (1977) (ed.) *Scale effects in animal locomotion*. London: Academic Press.

Pennycuick, C. J. (1968) Power requirements for horizontal flight in the pigeon *Columba livia*. *Journal of Experimental Biology*, **49**, 509–26.

Pennycuick, C. J. (1974) *Handy matrices*. London: Edward Arnold.

Prandtl, L. (1952) *Essentials of fluid dynamics*. London & Glasgow: Blackie.

Prandtl, L. & Tietgens, O. G. (1934) *Applied hydro- and aerodynamics*. New York: McGraw Hill.

Pyatetskiy, V. E. (1970) Hydrodynamic characteristics of swimming of some fast marine fish. *Bionika*, **1970**, 20–27.

Rayner, J. M. V. (1979) A new approach to animal flight mechanics. *Journal of Experimental Biology*, **80**, 17–54.

Rayner, M. D. & Keenan, M. J. (1967) Role of red and white muscles in the swimming of skipjack tuna. *Nature, London*, **214**, 392–3.

Richardson, E. G. (1936) The physical aspects of fish locomotion. *Journal of Experimental Biology*, **13**, 63–74.

Ripkin, J. F. & Pilch, M. (1964) Non-newtonian pipe friction studies with various dilute polymer water solutions. *St Anthony Falls Hydraulic Laboratory Report*, number 11.

Rosen, M. W. (1959) Water flow about a swimming fish. *United States Naval Ordinance Test Station Technical Publication*, number 2298, 1–96.

Rosen, M. W. (1961) Experiments with swimming fish and dolphins. *American Society of Mechanical Engineers*, publication 61–WA–203, 1–11.

Rosen, M. W. & Cornfield, N. E. (1971) Fluid friction of fish slimes. *Nature, London*, **234**, 49–50.

Rosenthal, H. (1968) Schwimmverhalten und schwimmgeschwindigkeit bei den larven des herrings *Clupea harengus*. *Helgoländer wissenschaftliche Meeresuntersuchungen*, **18**, 453–86.

Rosenthal, H. & Hempel, G. (1969) Experimental studies in feeding and food requirements of herring larvae. In *Symposium on Marine Food Chains*, ed. J. H. Steele, pp. 344–64. Edinburgh: Oliver & Boyd.

Schlichting, H. (1952) *Boundary layer theory*. New York: McGraw-Hill.

Shann, E. W. (1914) On the nature of the lateral muscle in the teleostei. *Proceedings of the Zoological Society of London* (*1914*), 195–215.

Shapiro, A. H. (1964) *Shape and flow*. London: Heinemann.

Shoulejkin, W. (1929) Aerodynamics of the flying fish. *International Review of Hydrobiology and Hydrogeography*, **22**, 102–110.

Simmons, J. R. (1970) The direction of the thrust produced by the heterocercal tail of two dissimilar elasmobranchs: the Port Jackson shark *Heterodontus portusjacksoni* (Meyer) and the Piked Dogfish, *Squalus megalops* (Mackleay). *Journal of Experimental Biology*, **52**, 95–107.

Slijper, E. J. (1958) *Walvissen*. Amsterdam: D. B. Centens, Vitgevers–Maatschappij.

Slijper, E. J. (1961) Locomotion and locomotory organs in whales and dolphins (Cetacea). In *Vertebrate locomotion*, Symposium of the Zoological Society of London, **5**, pp. 77–94.

Smit, H. (1965) Some experiments on the oxygen consumption of Goldfish (*Carassius auratus* L) in relation to swimming speed. *Canadian Journal of Zoology*, **43**, 623–33.

Smit, H., Amelink-Koustaal, J. M. & Vijverberg, J. (1971) Oxygen consumption and efficiency of swimming goldfish. *Comparative Biochemistry and Physiology*, **39**, 1–28.

Sterba, G. (1962) *Freshwater fishes of the world*. London: Vista.

Sundnes, G. (1963) Energy metabolism and migration of fish. *Northwest Atlantic Environmental Symposium, Special Publication*, number 6, pp. 743–46.

Taylor, G. I. (1952) Analysis of the swimming of long narrow animals. *Proceedings of the Royal Society* (*A*), **214**, 158–83.

Thomson, K. S. (1976) On the heterocercal tail in sharks. *Palaeobiology*, **2**, 19–38.

Thomson, K. S. & Simanek, D. E. (1978) Body form and locomotion in sharks. *American Zoologist*, **17**, 343–54.

Tritton, D. J. (1977) *Physical fluid dynamics*. London: van Nostrand Reinhold.

Tucker, V. A. (1973) Bird metabolism during flight: evaluation of a theory. *Journal of Experimental Biology*, **58**, 689–709.

van der Stelt, A. (1968) *Spiermechanica en myotoombouw bij vissen*. Unpublished Ph.D. Thesis, University of Amsterdam.

Videler, J. J. (1975) On the interrelationships between morphology and movement in the tail of the cichlid fish *Tilapia nilotica* (L). *Netherlands Journal of Zoology*, **25**, 143–94.

Videler, J. J. (1981) Swimming movements, body structure and propulsion in cod *Gadus morhua*. In *Vertebrate locomotion*, Symposium of the Zoological Society of London, **48**, 1–27.

Videler, J. J. & Wardle, C. S. (1978) New kinematic data from high speed cine film recordings of swimming cod (*Gadus morhua*). *Netherlands Journal of Zoology*, **28**, 465–84.

Vlymen, W. J. (1974) Swimming energetics of the larval anchovy, *Engraulis mordax*. *Fishery Bulletin*, **72**, 885–99.

von Holst, E. & von Kuchemann, D. (1942) Biological and aerodynamic problems of animal flight. *Journal of the Royal Aeronautical Society*, **46**, 44–54.

von Mises, R. (1959) *Theory of flight*. New York: Dover Publications.

Walters, V. (1962) Body form and swimming performance in scombrid fishes. *American Zoologist*, **2**, 143–9.

Walters, V. (1963) The trachypterid integument and an hypothesis on its hydrodynamic function. *Copeia*, 1963, 260–70.

Wardle, C. S. (1975) Limit to fish swimming speed. *Nature, London*, **225**, 725–7.

Wardle, C. S. (1977) Effect of size on swimming speeds of fish. In *Scale effects in animal locomotion*, ed. T. J. Pedley, pp. 299–313. New York & London: Academic Press.

Wardle, C. S. & Reid, A. (1977) The application of large amplitude elongated body theory to measure swimming power in fish. In *Fisheries mathematics*, ed. J. H. Steele, pp. 171–91. London: Academic Press.

Wardle, C. S. & Videler, J. J. (1980) Fish swimming. In *Aspects of animal movement*. eds H. Y. Elder & E. R. Trueman, pp. 125–50. Cambridge University Press.

Ware, D. M. (1975) Growth, metabolism and optimal swimming speed of a pelagic fish. *Journal of the Fisheries Research Board of Canada*, **32**, 33–41.

Ware, D. M. (1978) Bioenergetics of pelagic fish: theoretical changes in swimming speed and relation with body size. *Journal of the Fisheries Research Board of Canada*, **35**, 220–28.

Webb, P. W. (1970) Some aspects of the energetics of swimming fish with special reference to the rainbow trout. Unpublished Ph.D. Thesis, University of Bristol.

Webb, P. W. (1971*a*) The swimming energetics of trout I. Thrust and power at cruising speeds. *Journal of Experimental Biology*, **55**, 489–520.

Webb, P. W. (1971*b*) The swimming energetics of trout II. Oxygen consumption and swimming efficiency. *Journal of Experimental Biology*, **55**, 521–40.

Webb, P. W. (1973*a*) Effects of partial caudal fin amputation on the kinematics and metabolic rate of underyearling salmon (*Oncorhynchus nerka*) at steady swimming speeds. *Journal of Experimental Biology*, **59**, 565–81.

Webb, P. W. (1973*b*). Kinematics of pectoral fin propulsion in *Cymatogaster aggregata*. *Journal of Experimental Biology*, **59**, 697–710.

Webb, P. W. (1974) Efficiency of pectoral fin propulsion of *Cymatogaster aggregata*. In *Swimming and flying in nature*, vol. 2, eds T. Y. Wu, C. J. Brokaw & C. Brennen, pp. 573–84. New York: Plenum.

Webb, P. W. (1975*a*) Hydrodynamics and energetics of fish propulsion. *Bulletin of the Fisheries Research Board of Canada*, number 190, 1–159.

Webb, P. W. (1975*b*) Acceleration performance of rainbow trout *Salmo gairdneri* (Richardson) and green sunfish *Lepomis cyanellus* (Rafinesque). *Journal of Experimental Biology*, **63**, 451–65.

Webb, P. W. (1976) The effect of size on the fast-start performance of rainbow trout, *Salmo gairdneri* and a consideration of piscivorous predator–prey interactions. *Journal of Experimental Biology*, **65**, 157–77.

Webb, P. W. (1977) Effects of partial fin amputation on fast-start performance of rainbow trout (*Salmo gairdneri*). *Journal of Experimental Biology*, **68**, 123–35.

Webb, P. W. (1978) Fast-start performance and body form in seven species of teleost fish. *Journal of Experimental Biology*, **74**, 211–26.

Webb, P. W. (1982) Locomotor patterns in the evolution of actinopterygian fishes. *American Zoologist*, **22**, 329–42.

Webb, P. W. & Keyes, R. S. (1981) Division of labor in the median fins of the dolphin fish, *Coryphaena lippurus*. *Copeia*, 1981, 901–4.

Webb, P. W. & Skadsen, J. M. (1980) Strike tactics of *Esox*. *Canadian Journal of Zoology*, **58**, 1462–9.

Weihs, D. (1972) A hydromechanical analysis of fish turning manoeuvres. *Proceedings of the Royal Society of London (B)*, **182**, 59–72.

Weihs, D. (1973*a*) The mechanism of rapid starting of slender fish. *Biorheology*, **10**, 343–50.

Weihs, D. (1973*b*) Optimal fish cruising speed. *Nature, London*, **245**, 48–50.

Weihs, D. (1973*c*) Mechanically efficient swimming techniques for fish with negative buoyancy. *Journal of Marine Research*, **31**, 194–209.

Weihs, D. (1974*a*) Energetic advantages of burst swimming of fish. *Journal of Theoretical Biology*, **48**, 215–29.

Weihs, D. (1974*b*) Some hydromechanical aspects of fish schooling. In *Swimming and flying in nature*, vol. 2, eds T. Y. Wu, C. J. Brokaw & C. Brennen, pp. 203–18. New York: Plenum.

Weihs, D. (1975) An optimal swimming speed of fish based on feeding efficiency. *Israel Journal of Technology*, **13**, 163–7.

Weihs, D. (1977) Effects of size on sustained swimming speeds of aquatic organisms. In *Scale effects in animal locomotion*, ed. T. J. Pedley, pp. 333–38. London: Academic Press.

Weihs, D. (1978) Tidal stream transport as an efficient method for migration. *Journal du Conseil*, **38**, 92–9.

Weihs, D. (1980) Energetic significance of changes in swimming modes during growth of larval anchovy, *Engraulis mordax*. *Fishery Bulletin*, **77**, 597–604.

Weis-Fogh, T. (1973) Quick estimates of flight fitness in hovering animals, including novel mechanisms for lift production. *Journal of Experimental Biology*, **59**, 169–230.

Weis-Fogh, T. & Alexander, R. McN. (1977) The sustained power output from striated muscle. In *Scale effects in animal locomotion*, ed. T. J. Pedley, pp. 511–25. London: Academic Press.

White, D. C. S. (1977) Muscle mechanics. In *Mechanics and energetics of animal locomotion*, eds R. McN. Alexander & G. Goldspink, pp. 23–56. London: Chapman & Hall.

Willemse, J. J. (1966) Functional anatomy of the myosepta in fish. *Proceedings, Koninklijke (Nederlandse) Akademie van Wetenschappen, (Series C)*, **69**, 58–63.

Wu, T. Y. (1961) Swimming of a waving plate. *Journal of Fluid Mechanics*, **10**, 321–44.

Wu, T. Y. (1971*a*) Hydromechanics of swimming propulsion. Part 1. Swimming of a two-dimensional flexible plate at variable forward speeds in an inviscid fluid. *Journal of Fluid Mechanics*, **46**, 337–55.

Wu, T. Y. (1971*b*) Hydromechanics of swimming propulsion. Part 2. Some optimum shape problems. *Journal of Fluid Mechanics*, **46**, 521–44.

Wu, T. Y. (1971*c*) Hydromechanics of swimming propulsion. Part 3. Swimming and optimum movements of a slender fish with side fins. *Journal of Fluid Mechanics*, **46**, 545–68.

Wu, T. Y. (1971*d*) Hydromechanics of swimming fishes and cetaceans. *Advances in Applied Mathematics*, **11**, 1–63.

Wu, T. Y., Brokaw, C. J. & Brennan, C. (1974) (eds) *Swimming and flying in nature*, vol. 2. New York, Plenum.

Index